DRY CLIMATE GARDENING

Inspiring | Educating | Creating | Entertaining

Brimming with creative inspiration, how-to projects, and useful information to enrich your everyday life, quarto.com is a favorite destination for those pursuing their interests and passions.

First Published in 2023 by Cool Springs Press, an imprint of The Quarto Group,
100 Cummings Center, Suite 265-D, Beverly, MA 01915, USA.
T (978) 282-9590 F (978) 283-2742 Quarto.com

27 26 25 24 23   1 2 3 4 5

ISBN: 978-0-7603-7702-4

Digital edition published in 2023
eISBN: 978-0-7603-7703-1

Library of Congress Cataloging-in-Publication Data

Names: Johnson, Noelle (Horticulturist), author.
Title: Dry climate gardening : growing beautiful, sustainable gardens in
  low-water conditions / Noelle Johnson, the AZ Plant Lady.
Other titles: Growing beautiful, sustainable gardens in low-water conditions
Description: Beverly, MA, USA : Cool Springs Press, 2023. | Includes index.
  | Summary: "Dry Climate Gardening shows gardeners in arid climates how
  to create a low-water landscape that's vibrant and colorful with lots of
  texture and interest"-- Provided by publisher.
Identifiers: LCCN 2022032994 (print) | LCCN 2022032995 (ebook) | ISBN
  9780760377024 (trade paperback) | ISBN 9780760377031 (ebook)
Subjects: LCSH: Desert gardening. | Desert plants. | Handbooks and manuals.
Classification: LCC SB427.5 .J65 2023 (print) | LCC SB427.5 (ebook) | DDC
  635.9/525--dc23/eng/20220822
LC record available at https://lccn.loc.gov/2022032994
LC ebook record available at https://lccn.loc.gov/2022032995

Design: Bad People Good Things LLC
Photography: See Photography Credits on page 197.
Illustration: Holly Neel

Printed in Singapore

APRIL 2023

Cover Photo: The lavender blooms of 'Lynn's Legacy' Texas sage (*Leucophyllum langmaniae*) add color and softness amid the spiky shapes of Weber agave (*Agave weberi*) and blue yucca (*Yucca rigida*) at the Desert Botanical Garden, in Phoenix, Arizona.

DRY CLIMATE GARDENING

GROWING BEAUTIFUL,
SUSTAINABLE GARDENS
IN LOW-WATER CONDITIONS

NOELLE JOHNSON *The AZ Plant Lady*

COOL
SPRINGS
PRESS

FOR THOSE WHO
YEARN FOR A
BEAUTIFUL GARDEN
IN THE MIDST OF
A DRY LAND . . .

Counter-clockwise from top left: Pink and white globe mallow flowers *Sphaeralcea ambigua* • Pink fairy duster *Calliandra eriophylla* • Variegated smooth edge agave *Agave desmettiana* 'Variegata' and autumn sage *Salvia greggii* • Bells of Fire™ Tecoma *Tecoma* x 'Bells of Fire' • Purple lilac vine *Hardenbergia violaceae* • Yucca flower and orange *Tecoma* x 'Orange Jubilee' • *Artichoke agave* Agave parryi var. truncata with fallen palo verde flowers

CONTENTS

|||||||||||||||||||||||||||||||||||||||||||||||||||||||||||||||||

INTRODUCTION

WELCOME TO GARDENING IN A DRY CLIMATE

Does the thought of creating a garden in a dry climate feel intimidating to you? It is not surprising in a region where scarce rain and low humidity are the norms. However, when you throw in sizzling hot summer temperatures, that adds another element of stress for plants and you, the gardener. It can be overwhelming as our seasons are different from everywhere else. You don't have to look far to hear news of changing weather patterns that translate into increasing drought conditions and record-breaking temperatures. Perhaps you have even seen the effects of climate change in your garden, with established plants showing new signs of stress.

So, how do you deal with these increasing stresses without settling for a dull garden with few plants? The answer is surprisingly simple—we need to create resilient gardens that can handle these challenges of increasing temperatures and drought. Gardening in a climate of extremes doesn't have to be difficult, but we tend to make it so by repeating common mistakes. These mistakes range from choosing the wrong plants for hot, dry climates that are ill-suited for the challenging conditions endemic to our regions. Surprisingly, people also water their plants incorrectly, often irrigating too often and not deeply enough. And then, there is a tendency to over-maintain plants by excessive pruning and applying fertilizer when they don't need it in many cases.

The regular gardening rules in regions with mild summers and more rainfall don't work well here and can frustrate you. But whether you are brand new to gardening or an experienced gardener from another climate, you can garden successfully in an arid region. The key to a beautiful outdoor space in a dry environment is to avoid common mistakes and embark on a journey of learning how to create, grow, and maintain a garden that thrives in a hot, arid climate.

I invite you to think of your perception of a dry climate garden. Given our challenging conditions, your vision may be limited to a landscape filled with only cactus and rocks. However, I'm here to tell you that so much more is possible within a dry climate! You can have a garden filled with flowering shrubs and groundcovers or one filled with unique shapes of cacti and succulents, or a combination of both. The key to success in a hot, dry land is to learn a new set of guidelines and embrace arid-adapted plants to achieve your goal of a beautiful outdoor space that will shrug off the stress of low humidity, drought, and intense heat.

A colorful yet resilient low-desert garden planted with saguaro cacti (*Carnegiea gigantea*), senita cacti (*Pachycereus schottii*), and Bells of Fire™ Tecoma (*Tecoma stans* 'Bells of Fire').

The spiky leaves of Caribbean agave (*Agave angustifolia*) add great texture contrast alongside rock verbena (*Glandularia pulchella*).

Golden barrel cacti (*Echinocactus grusonii*) and organ pipe cactus (*Stenocereus thurberi*) create a lovely visual pairing with purple and white trailing lantana (*Lantana montevidensis*).

CHALLENGES OF GROWING PLANTS IN A HOT, ARID CLIMATE

Whatever region you live in, there are challenges to growing plants, and dry climates are no exception. Plants are subject to temperatures ranging from intensely hot, 110°F-plus (43°C) in summer, to freezing winter temperatures. Add to that extreme dryness, sporadic rainfall, and periods of drought, it's no surprise that people struggle to create an attractive garden that thrives despite these challenges.

I'm here to tell you that there are plants that survive and thrive in an arid region while adding welcome beauty to your home. So, whether you find yourself longing for lush green plants and a colorful garden, one filled with cacti and succulents, or a combination of both, you have come to the right place! Within the pages of

this book, you will learn what plants do best in hot, dry regions, along with tips on watering and plant care to ensure your garden will thrive and add beauty outside your home. And it is easier than you think!

I will show you why the regular gardening rules from regions with mild summers and more rainfall don't work well here. Beginning with the cornerstone of a healthy garden, you'll learn why shade is highly prized, choose plants that will thrive in arid regions, learn the right way to care for plants, and avoid common pitfalls. A dry climate garden doesn't have to be boring, and I'll show you how to design a beautiful outdoor space that will compliment your home. At a loss as to what plants to use in your garden? Toward the end of this book, I've created plant profiles of my favorite trees,

I enjoy spending time in my garden and gravitate toward color and texture, like planting succulents in bright blue pots.

shrubs, groundcovers, cacti, and succulents for dry climate gardens and specialized plant lists for specific conditions.

MY DRY CLIMATE GARDENING JOURNEY

I have lived in dry climate regions my entire life. From a childhood spent in the semi-arid, Mediterranean climate of Southern California to my adult life in the low desert of Arizona, drought, heat, and low humidity are challenges I have always dealt with in the garden. As a horticulturist and landscape consultant, I have been helping people just like you learn how to create, grow, and maintain beautiful outdoor spaces that thrive in dry climates for over twenty-five years.

However, I didn't grow up with a green thumb. I actually killed all the plants in my first desert garden. I made so many mistakes because I followed gardening rules and guidelines that were not tailored for a hot, dry climate. I remember feeling so frustrated by my efforts resulting in a yard filled with dead plants that I had planted less than a year before. Little did I know at the time that this failure was the catalyst that would lead me onto a rich and rewarding career.

It began as a quest to figure out how to garden the "right" way, working within my desert climate's hot, dry norms instead of fighting them. My search eventually led to me going back to school for a horticulture degree. I stopped seeing my climate as a liability, and instead of a place where I could achieve garden success if I embraced the correct guidelines and strategies for this land of dryness, heat, and intense sun.

If I had to sum up in one sentence what I learned in over twenty-five years as a horticulturist, it would be this: Gardening in a hot, dry climate isn't hard, but it is different. I am so excited for you to pick up this book because you are taking the first step toward embracing and exploring the wonderful world of dry climate gardening and all that is possible!

If you walk through my garden today, you will see a landscape filled with vibrant shades of green and colorful flowers from shrubs, ground-covers, and vines. Cacti and other succulents add wonderful texture contrast and unique interest, while trees provide welcome filtered shade. While my garden is beautiful, there is a story behind it of plants that have died and my mistakes. Any successful garden shares this same background story. But these mistakes and plant failures have enabled me to learn what does best and, more importantly, what

The desert contains a surprising amount of diversity in the variety of plants that flourish despite challenging conditions.

Deserts can undergo wide swings in temperature from winter to summer. Winter snow covers cacti and other native plants in the foothills of Phoenix, Arizona.

doesn't. Gardening in a dry climate involves learning some different rules. However, whether you are brand new to gardening or are an experienced gardener from another type of climate, you can garden successfully in arid regions.

WHAT IS A DRY CLIMATE?

How do you know if you live in one? First, let's break it down even more by exploring specific characteristics that all types of dry climates share. Then I will help you identify the differences between arid and semi-arid regions to help you determine which one you live in.

Characteristics of dry climate regions:
- low precipitation and sporadic rain events
- dry air or low humidity present throughout most of the year
- big swings in temperature from intensely hot summers to mild or cold winters

While these characteristics are present in all dry climates at varying levels, for this book, I am talking about how to garden in semi-arid and arid areas that experience hot summers. Specifically, there are two dry climates—arid (desert) and semi-arid (also referred to as Mediterranean).

DESERT (ARID)

Intensely hot summers and mild to cold winters are the hallmarks of deserts we are addressing in this book, which encompass approximately 14 percent of the world's landmass. The severity of heat and cold are tied to the altitude of a specific desert region. Lower elevation deserts, below 2,000 feet (610 m) in altitude, are hotter in summer than higher elevations. Winters can range from frost-free to a handful of days below freezing in the low desert. Mid to high altitude deserts, 2,500 to 6,000 feet (610–1829 m), experience steadily colder winters as you go higher, dropping below 0°F (-18°C) in high desert regions. In most deserts, summer days are usually 100°F (38°C) or more and are getting hotter with climate change.

In the desert, the humidity level during the summer can go down to the single digits, which puts severe stress on ill-adapted plants. With their lower temperatures and sporadic rainfall, winter months have higher levels of water vapor in the air. However, humidity levels seldom reach above 40 percent and usually remain in the 30 percent range.

Low precipitation is the hallmark of all deserts, averaging less than 14 inches (35.6 cm) annually and often much less depending on the specific desert region. In most deserts, the majority of

A grove of oaks and native grasses grows in Central California where winters are mild and summers are warm to hot and dry.

Cacti, succulents, and flowering plants combine to create a beautiful display with spiky yucca and prickly pear cacti with flowering shrubs like 'Lynn's Legacy' Texas sage (*Leucophyllum langmaniae* 'Lynn's Legacy')

rainfall arrives in the winter months. As a result, summers are usually dry, but some fortunate desert regions experience two summers. The first is a dry *fore-summer* that spans the first half of summer, and the second part is a *wet (monsoon) summer*. During the second half of summer, seasonal wind shifts bring tropical moisture to thirsty deserts. The Southwestern region of the United States experiences summer rain during the monsoon season.

MEDITERRANEAN (SEMI-ARID)

An apt description of semi-arid regions is that of a gentler version of the desert regarding rain, humidity, and temperature challenges. They are a dry(ish) climate that transitions between deserts and more temperate climates. They experience milder temperatures than deserts, with summer temps ranging from 70–90°F (21–32°C). There are days above 100°F (38°C) in the summer, but they tend to be sporadic. Temperatures in winter are relatively mild, ranging in the 50s (10°C), and freezes are rare.

Higher humidity levels allow more trees and shrubs to grow with natural rainfall than deserts. Fog can develop in semi-arid climates, particularly near the coast, enabling plants to grow despite low rain amounts. Hence, they are often referred to as Mediterranean climates as they share the general climatic characteristics of regions bordering the Mediterranean Sea.

Annual rainfall amounts seldom reach more than 20 inches (51 cm), and most of it falls in winter. Periods of drought do occur and are happening with greater frequency. Summers are warm and dry, while winters can be wet when rainfall occurs in average amounts.

GET READY TO CREATE A BEAUTIFUL AND RESILIENT DRY CLIMATE GARDEN

I never cease to be amazed at the sheer amount of plants that do well in the challenging conditions present in a dry climate. Sadly, people think of dry climate gardens as landscapes filled with rocks, cacti, and a distinct lack of color. Yet, nothing is further from the truth. Arid-adapted plants come in all colors of the rainbow, with many having brilliant tones that provide colorful impact. And instead of most plants being the same shade of saturated green you see in many regions, the dry climate garden is one with many different shades of green ranging from gray-green or sage greens to deep green tones. As a result, you can have a beautiful landscape around your home that will bring you joy throughout all seasons with less upkeep than you might imagine. I am excited to embark on this garden journey with you!

DRY CLIMATE REGIONS |||||||||||||||||||||||||||||||||||||||||||||||||||||||||||||||||||||||||||||||||||||

Arid Regions

Semi-Arid Regions

CHAPTER 1

CHOOSING THE RIGHT PLANTS FOR HOT, ARID REGIONS

Selecting suitable plants for your dry climate garden is the most critical element to an attractive and resilient landscape that will thrive amidst hot summers, low humidity, and even cold winters. Unfortunately, countless people struggle to grow plants that are not well-suited to handle these stresses. As a result, they spend a lot of time and money only to become frustrated with plants that struggle to perform well, which creates a lackluster landscape. So, the first step is to let go of plants that struggle and move forward to embrace the beauty and uniqueness of arid-adapted plants.

NATIVE PLANTS

The first type of plants to consider are those native to your region. These grow in natural areas throughout your town or city, including nearby areas. Take a drive outside your city limits and look at what grows naturally. Native plants are the easiest and most low-maintenance option for wherever you live. Natives usually rely on rainfall alone in dry climate settings, making them ideal for the ultimate low-maintenance and low-water garden. However, native plants will need supplemental water when first planted to become established with a robust root system and in times of drought. Another benefit is that they can withstand the low humidity, high summer temperatures, and winter lows. Finally, when it comes to insect pests, native plants are rarely bothered by them as nonnatives sometimes are.

A low-maintenance desert landscape is filled with plants native to the Southwestern United States, including buckhorn cholla cacti (*Cylindropuntia acanthocarpa*), brittlebush (*Encelia farinosa*), compass barrel cactus (*Ferocactus wislizeni*), chuparosa (*Justicia californica*), foothills palo verde (*Parkinsonia microphylla*), and firecracker penstemon (*Penstemon eatonii*).

A mixture of native shrubs and succulents thrive in a low desert garden, including black spine agave (*Agave macroacantha*), Parry's agave (*Agave parryi*), desert milkweed (*Asclepias subulata*), Baja fairy duster (*Calliandra californica*), and goldeneye (*Viguiera parishii*).

Another positive aspect of using native plants is that they attract pollinators such as birds, butterflies, and bees to your garden so you can observe them up close. Native trees, shrubs, and groundcovers generally look better in a garden than when growing in their natural environment. While they usually thrive in your garden with little to no attention, you can provide supplemental water in times of drought, or sporadic pruning, if needed.

A common assertion is that using native plants solely is the only way to create a healthy, flourishing garden—and yes, you can create a beautiful outdoor space using only native trees, shrubs, groundcovers, cacti, and other succulents. But, you certainly can go beyond planting natives only by incorporating arid-adapted plants from different parts of the world.

ARID-ADAPTED PLANTS

So, what are arid-adapted plants or plants adapted to dry climate gardens? These are plants from other regions with similar temperatures, sun, rainfall, and humidity conditions. When discussing arid-adapted plants, we look at plants native to other dry climate regions, such as those with desert or Mediterranean climates.

Arid-adapted natives from different continents, Mexican honeysuckle (*Justicia spicigera*) and shrubby germander (*Teucrium fruticans* 'Azurea') make a great pair along a garden path.

DITCH YOUR HOMETOWN PLANTS

A common mistake people make when they move to a dry climate is trying to grow their favorite plants from their hometown that are adapted to higher humidity, more rainfall, and milder summers. Growing plants in a dry climate presents challenges in low humidity, intense sunlight, and hot temperatures, making it difficult for plants that aren't equipped to deal with them. When you ignore the needs of plants, you run into trouble, spending more money on replacing plants and extra water for irrigation. Instead, embrace the incredible variety of plants that will thrive where you now live.

Areas include the desert Southwest of the United States, areas around the Mediterranean Sea, South Africa, the west coast of South America, and the Middle East.

I've seen landscapes in the Middle East filled with plants native to the deserts of the United States, which thrive a world away in my own Arizona garden. I also have several flowering shrubs native to Australia that have provided me with beauty for over twenty years. The world interconnects through the plants in our gardens.

INCORPORATING PLANTS FROM OTHER ARID REGIONS IN YOUR GARDEN

Look at what type of arid climate you have when selecting plants for your garden. For desert dwellers who regularly deal with extreme heat and dryness, it is essential to choose plants native to desert regions to ensure they will thrive. Yes, some Mediterranean native plants will survive in the desert. Still, not all can handle the blistering summer, so research plants ahead of time from other regions to make sure they can handle desert conditions. The same strategy is true if you live in a semi-arid climate with more moderate temperatures than the desert—select

plants native to similar environments to yours. Many trees, shrubs, groundcovers, and succulents native to the desert regions will also do well in a semi-arid Mediterranean climate.

TYPES OF PLANTS THAT STRUGGLE IN DRY CLIMATES THAT YOU SHOULD AVOID

While many plants do well in dry climates, quite a few perform poorly. These plants have high water requirements and lack the adaptations to limit water loss through the leaves. In addition, plants from areas with higher rainfall struggle in the alkaline soil prevalent in arid regions as they tend to prefer acidic soils. Finally, not all plants can handle hot summers and low humidity.

With climate change, drought is more prevalent, and temperatures are getting hotter. Plants that are ill-adapted have a more challenging time managing those stresses, and it shows in

Tecoma hybrid 'Bells of Fire'
BELLS OF FIRE™
fast growing bushy semi-deciduous broadleaf shrub
height 3 to 5 feet - spread 3 to 4 feet
full sun - hardy to about 10 degrees f
does well in containers
red flowers spring thru fall

30% organic mulch
low water use

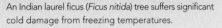

An Indian laurel ficus (*Ficus nitida*) tree suffers significant cold damage from freezing temperatures.

Look for plant labeling that is specific to your climate, which you can often find at local nurseries.

stunted growth, sunburned foliage, and a failure to thrive. Therefore, we need to plant gardens filled with resilient plants that can endure these extremes more than ever.

PLANTING ZONES AND WHY MINIMUM COLD TEMPERATURE ISN'T ENOUGH

Wherever you live, your region is classified within a planting zone. Different countries have their own plant zone classifications, but almost all base the zones on the minimum cold temperature that the region experiences in winter. Plants have a base cold temperature that they can tolerate, and any colder than that, they won't survive. For example, some plants native to tropical regions are frost-tender and cannot handle freezing temperatures. In contrast, extremely cold-hardy plants can endure temperatures below 0°F (-18°C). When you look at a plant label or description, it likely includes a planting zone classification to tell you if it is suitable for the winter cold your area experiences.

However, in climates that experience hot summers, minimum cold temperatures are just part of what you need to know. For example, the low desert regions in Arizona are in the same cold hardiness planting zone as coastal northwestern California and its redwood forests. However, both areas have very different climatic conditions—especially in summer. So, if you only use planting zones based on cold hardiness as your only metric for selecting plants for your dry climate garden, you are missing a big part of the puzzle. In addition to the minimum cold temperature, we also need to know how much heat and aridity a plant can tolerate. You can find some heat zone maps if you are fortunate, but sadly, these aren't widely available worldwide and are rarely included on a plant label. Thankfully, there are other ways we can determine what plants can handle our hot summers, which I will mention later in this chapter.

DECIPHERING PLANT LABELS

You've seen them—a plant label in the form of a little plastic tag inserted in the soil by your new plant or a sticker on the nursery container. Of course, like any savvy consumer, you read the label before purchasing your plant to make sure you meet the plant's requirements in your garden. However, the information on a plant label isn't always accurate for arid regions—particularly those in the desert.

First, let's look at what information from the plant label is helpful: the plant's mature size and the common and botanical names. We need to know how large a plant will grow to allow enough room in our garden. You are likely to be more comfortable using common plant names than botanical names, not to mention trying to pronounce them. However, common names aren't always reliable when identifying plants, as several completely different plants can share the same common name. So, botanical (also known as scientific) names are crucial as there is only one plant per name.

Now, let us look at what information on a plant label doesn't always apply to dry climate gardens with intense heat and sun. First, the bloom season may differ slightly, with flowers appearing earlier or later than stated on the label. But the most common misleading information on a label has to do with recommended sun exposure. Often, these miss the mark, and plants whose label stated "full sun" may struggle to grow and may even die in hot, arid climates if planted in full sun. This misinformation exists as the grower doesn't factor in the intensity of the sun's rays coupled with the fact that heat in a desert zone is much higher than in other regions, including Mediterranean climates.

Autumn sage (*Salvia greggii*) thrives under the filtered sunlight of a palo verde tree as it cannot tolerate the intensity of the summer sun in low desert regions, despite its label stating full sun.

WHY IS SUN AND SHADE EXPOSURE SUCH A BIG DEAL?

Every plant has specific requirements to grow well and add beauty to your garden. Ignore these needs, and you will have a plant that struggles, needs more attention, more water, more fertilizer, etc. Sadly, not considering a plant's sun or shade requirements is one of the most common mistakes I see people make in the garden.

HOW THE SUN AFFECTS PLANTS

If you've ever added a plant only to watch it struggle to survive despite keeping it well-watered, it usually has to do with the intensity of the sun and the low humidity in dry climates. To understand why the amount of sun is such a big deal, let's talk about how the sun's intensity relates to water loss in plants. It comes down to the fact that the sun pulls out water from a plant's leaves in a process known as transpiration: The more sun, the more potential water loss from the plant. In addition, more water loss from the leaves occurs in low

Gopher plant (*euphorbia rigida*) and angelita daisy (*Tetraneuris acaulis*) are well-adapted to growing in the corner next to a driveway and street despite the intense sun and reflected heat from the pavement.

humidity due to the dry air. In extreme cases, leaves wilt and become sunburned when too much water loss occurs.

A plant put in an area with full sun that isn't adapted to handle the stresses of a hot, dry climate will struggle because it doesn't have the adaptations to help prevent excess water loss through its leaves. For example, one of the reasons that succulents are so prevalent in arid regions is that they have a unique way of preventing water loss during the day. Succulents do this by closing their stomata in the daytime and opening them at night instead. Stomata are tiny openings on the leaves of plants that allow them to breathe, and it's where water loss occurs. Other types of plants such as trees, shrubs, and groundcovers open their stomata during the day.

HOW PLANTS ADAPT TO DRY HEAT AND FULL, REFLECTED SUN

If your landscape has areas that bake in the sun all day, the good news is that there are many types of plants that will survive these challenging conditions and thrive. With trees, shrubs, vines, and groundcovers, it all comes down to the ability of a particular plant to reduce the amount of water lost through its leaves during the day.

Many plants native to areas with dry climates have adaptations to survive and flourish. One characteristic you will often see with plants is smaller leaves. The reduced leaf surface means that there is less area for water loss. Additionally, many trees and plants have green stems or trunks, which allow them to shed their leaves during times of drought, yet they can still photosynthesize—dry climate plants are so cool! Other adaptions that plants have to prevent water loss include a thick waxy cuticle over the leaf and the ability to move their leaves vertically to avoid the full brunt of the sun's rays.

This native Australian shrub, *Hypocalymma angustifolium* 'Coconut Ice', grows naturally in Western Australia. Its narrow leaves limit the amount of water lost through its leaves, enabling it to thrive in dry climates.

HOW PLANTS PREVENT WATER LOSS

An interesting adaptation many arid-adapted plants have that helps them prevent the loss of water are tiny hairs that coat the leaf's surface, giving them their subtle grayish hue. The hairs make the leaves soft to the touch and help reflect the sun's rays and inhibit water loss to the atmosphere.

Tiny hairs cover the leaf of a flowering Texas sage (*Leucophyllum frutescens*) that help to reflect the sun's rays and prevent excess water loss.

MICROCLIMATES

Microclimates are small areas around your house where sunlight, temperature, and even water accessibility are slightly different, which allow plants that otherwise may struggle within normal conditions found in arid and semi-arid gardens. Examples are landscapes filled with mature trees and groundcovers, which lower the temperatures while providing respite from the intensity of the unobstructed rays of the sun. Likewise, a courtyard or outdoor patio with an overhanging tree can provide welcome relief from the sun, making it easier to grow flowers, shrubs, or even vegetables in pots.

Filtered sunlight is an excellent compromise for plants that may grow in full sun in a more moderate climate but have a hard time in hot, dry weather. You can find areas of the dappled sun toward the outer reaches of tree branches that create partial sun.

The most common microclimate in arid regions is hot, reflected heat. Typically, these areas receive full sun and are found near walls and sidewalks, which absorb the heat from the sun and reflect it onto the plants. Plants can struggle in these hot zones unless they are native or have adaptations to handle the extreme heat and sun. However, in a colder arid climate, a garden with a warmer microclimate will enable you to grow plants that need some extra warmth.

Trees are the best way to add beauty to your outdoor space while creating more ideal growing conditions for plants with their shade. Shrubs, vines, and groundcovers are other plants that can also be useful to increase humidity of the garden while absorbing the heat of the day. A vine or tall shrub against a bare wall is an excellent way to absorb the heat from

WHAT IS YOUR NEIGHBOR GROWING?

My favorite tip for seeking out plants that will thrive in my garden is to observe what is growing in my neighborhood. You can do this when walking your dog or sitting in traffic. Look at the plants around you and note if they look healthy and attractive. Also, note what sun exposure they are growing in—full sun, partial sun, or shade. Then, take a picture of the plant and do some online sleuthing to determine what the plant is, or take the photo to your local nursery for help identifying it.

the sun and prevent it from being radiated back. It's helpful to think of plants as tools to reduce increasing temperatures and the heat-island effect growing in many urban areas.

STRATEGIES FOR FINDING PLANTS THAT WILL THRIVE IN AN ARID REGION

So, how can you tell what plants will do well in a dry climate? There are strategies to help you select the right plant. First, look at the characteristics that many arid-adapted plants have. These include succulent foliage, small leaves, and a grayish tint to the foliage color. Next, visit your local botanical garden to get a good idea of plants that do well where you live. While there, take special notice of the sun

exposure of where they are planted so you can replicate that at home. Take photos of plants you like and take a picture of the plant signage to help you identify the plant later. If no signage is available, an experienced nursery professional can help you identify the plant from the photo. Another option is to take a trip to your local nursery and walk through the aisles, observing the types of plants. You'll note that I said local nursery and not a big box store. The nursery professionals at a local nursery tend to be more knowledgeable about plants. While big box stores are a convenient source for plants, they often carry some plants that will do poorly in your region.

ONLINE PLANT SEARCHES THE RIGHT WAY

Online searches of plants can produce mixed results if not phrased correctly. Often, people only use the plant's name, and that pulls up information that may not be applicable to dry climate gardens. So instead, type the plant's name in your favorite search engine, and then add the name of what region you live in. For example, I live in the low desert so that I will type in "low desert" in the search bar next to the plant name. When you add your region to the search, if a plant does well where you live, it should show up in the results, and if it doesn't, that is a good indicator that it does poorly or it hasn't been grown widely enough to tell you if it will do well.

RELIABLE SOURCES FOR BUYING PLANTS

To create an attractive yet resilient garden that will perform well in a dry climate, we need to choose the "right" types of plants native to an arid or semi-arid region or known to do well. You might be surprised to learn that the most convenient place is the least reliable source. Here are four common sources for obtaining plants, along with the pros and cons of each:

LOCAL NURSERY OR GARDEN CENTER

These places cater to their local community and what their customers are looking for. Most will have certified nursery workers who can provide you with helpful information on how to grow specific plants and where they do best and will provide you with suggestions for certain areas of your garden. If you are looking for a particular plant that isn't in stock, they may be able to order it for you. Depending on the nursery, a limited warranty may or may not be available, so ask.

BIG BOX STORE

This is the most convenient choice for buying plants as there are many of them, so you do not have to drive far. Their prices tend to be the least expensive, and they do carry popular plant species for your particular region. However, many also stock plants that won't do well or are poorly adapted to growing in an arid or semi-arid climate. These plants are usually near the entrance, where their beautiful blooms entice customers into impulse buys. Sadly, the buyer heads home with their new plant only to be disappointed when it does poorly and thinks

Shopping cart filled with purple trailing lantana (*Lantana montevidensis*), Santa Rita prickly pear (*Opuntia santa-rita*), and globe mallow (*Sphaeralcea ambigua*) from a local botanic garden sale.

that it's their fault when in fact, the plant had no chance of doing well. Keep your receipt as many of these large stores will give you store credit if your plant dies. Big box store garden centers are a good option if you know what type of plant you are looking for, but avoid them if you aren't sure whether a plant will do well where you live.

Full-grown 'Rio Bravo' Texas sage (*Leucophyllum langmaniae* 'Rio Bravo') has reached its mature size three years after planting.

BOTANICAL GARDEN PLANT SALES

If you are fortunate to live near a botanical garden, this is a great place to observe plants to see which ones handle your climate best. They also can be a prime resource for purchasing plants. Plants sold at a local botanical garden are generally proven to do well where you live, and many are growing along the trails of the park itself. New plant varieties and tried and trusted species are offered, including some that are hard to find in a regular nursery. Plant sales are usually offered periodically, so plants may not be available for purchase all year. Look for plant signage that will indicate how to grow it specifically in your region. Proceeds from the sale often benefit the garden.

AVOID OVERPLANTING

The plants you purchase at the nursery are much smaller than they will be once planted and allowed to grow. This is because most plants at a nursery are young and in pots, which keeps them relatively small compared to their natural, mature size. However, once planted in the ground, roots have room to grow, resulting in the increased size of the plant and the number of branches and leaves. Some plants are slow growers while others have a fast growth rate, but all will grow significantly larger than what you see in the nursery.

LEARN THE MATURE SIZE OF PLANTS BEFORE YOU BUY

Once you know how large a plant will become, you can ensure that it will have enough room to grow within the site you have chosen for it. You can find information on how large a particular plant will reach at maturity from different sources: the nursery label, an online search, or a nursery professional. You'll need to know how wide the plant grows and how tall it will reach. Plant labels will typically show the width × height.

EXAMPLE: To determine how much room a shrub that reaches a mature size of approximately 6-feet wide (1.8 m) will need, take the width of 6 feet (1.8 m) and multiply 6 × 6 (1.8 × 1.8 m) = 36 sq. ft (4 sq. m) to come up with the amount of space to allow in your garden. I recommend adding at least another foot or two (30–61 cm) to provide for excess growth and make maintenance around the shrub easier.

So, before you choose a plant for a particular spot in your garden, you need to determine how much room exists for the plant to grow. A tape measure will become one of your most useful tools for planning your garden. While it may take several years for the plant to reach a mature size, preparing for its full size will prevent future problems with overcrowding.

If you follow the guidelines in this chapter when choosing plants, you are likely to have a garden with thriving plants. However, you also may experience some plants that fail to thrive or even die. It could be issues with exposure, soil condition, watering, or a plant that doesn't do well in a dry climate. I assure you that if you have plants die, it doesn't mean that you are a failure at gardening. Experienced gardeners look at a plant's death as a learning tool to help them determine what plants will do well and what adjustments may be needed with watering, sun exposure, etc. Gardening in a dry climate is a journey.

CHAPTER 2

PLANTING BASICS FOR DRY REGIONS

Shopping for plants is one of my favorite things to do. I love to walk through the aisles at my favorite nursery, looking at what's new, and I confess that I often come home with more plants than I plan to. Thankfully, I have a good-sized garden, so I usually find a spot for my new purchase. However, there is much more to planting than digging a hole and putting in a new plant. Sadly, most plant failures are due to mistakes that include ill-suited plants, wrong location, bad soil, and a poorly prepared hole. The good news is that these mistakes are easily avoided once you follow some basic guidelines.

First, let's talk about the process of planting. Even if you do everything right when you plant, it is a stressful experience for plants—you are moving a plant from its previous home at the nursery and placing it into an entirely new environment—different soil and different sun exposure. During this process, a plant goes through transplant shock, which is a period when it has to become accustomed to its new surroundings quickly. The length of time it takes a plant to get through this transition varies. Smaller plants like perennials and young shrubs go through it fairly quickly in weeks. The older the plant, such as shrubs in larger-sized containers, can take a few months to become acclimated to their new place in the garden. Trees can take at least a couple of years to become established.

During the transition from being newly planted to becoming established, plants first focus on growing a robust root system to support the leaves and branches. The roots gather water, oxygen, and nutrients for the plant so that the top part can grow. Without a sound root system, the rest of the plant will struggle. Plants also need to become used to the amount of sun or shade in their new location. This process can take a few weeks and can take even longer if planted in the summer. Temperature differences can also be a factor, especially when buying a plant from a grower in a different climate.

The success of these young plants rests on several factors, including time of year planted, the size of the hole, and whether or not you use amendments.

THE BEST TIME OF YEAR TO ADD NEW PLANTS

So how can we make the transition easier for new plants to become well established? While location and soil are important factors, a crucial element of success is *when* you add plants. In other words, we want to plant them when there are fewer stressful aspects present to help ensure their chance to grow and add beauty to our gardens. Let's explore the different seasons and their suitability for adding plants, ranked in order of preference.

FALL

The ideal time to plant is in autumn. When I tell people this, people are often surprised as they view fall as when gardening is slowing down. However, the milder temperatures in fall allow plants to get through transplant shock more easily without dealing with the stress that intense summer heat causes. But the fundamental reason that fall planting is favored is that it allows a plant three full seasons to grow a healthy root system, which is vital to helping a plant cope with the water demands of hot, dry summer. Cold-tender plants should be planted six to eight weeks before the first frost date that your region experiences. This timing will give plants some time to grow roots and harden off before winter cold arrives.

SPRING

Most people wait until spring to install new plants in their gardens, and it makes sense—temperatures are warm but not hot, and it's a time of year when the garden beckons us outdoors. Spring is the second-best season to add plants, and for frost-tender plants, it is the ideal time as you don't have to worry about young plants surviving winter cold. However, it's important to plant as early as possible in hot, dry climates, so your new plants have several weeks to grow roots and foliage before summer heat. So, how early can you plant in the spring without risking damage from a late freeze? The answer is simple: Find the last average frost date for your area. This information is available online. Once you know the average first frost date, you can plant any time after that, ideally, before temperatures rise above 95°F (35°C).

WINTER

In arid regions that experience frost-free winters, planting can continue through winter. In areas that experience occasional winter freezes at night but warm up in the day, you can add cold-hardy plants in this season. Native plants or those that are cold hardy are usually safe to plant in winter. If you live in an area that experiences mild winters, this is the best season to transplant existing plants to other areas within the garden when heat stress is absent.

SUMMER

It may not surprise you that summer is the least favorable time of year to add new plants. The intense heat and low humidity will stress plants already dealing with transplant shock. In addition, they have not had time to grow their roots into the soil and maximize their ability to absorb water. Summer heat and dry conditions use up most of a plant's energy to survive, and as a result, they can't focus on growing.

If you must plant in the summer, there are things you can do to minimize the stress on plants. First, dig the hole the day before and fill it with water, which will provide a reservoir of moisture for the new plant. Second, plant in the early evening to give your plant several hours to acclimate before the sun rises. Third, water in the new plant well after planting, with a hose on a slow trickle. Next, provide temporary shade covering using shade cloth until early fall. Finally, monitor your plant for signs of drought stress and water often, gradually weaning it to a regular watering schedule beginning in fall. Keep in mind that plants native to arid regions will handle summer planting much better than those from other areas.

PALM TREES LIKE IT WARM

Palm trees are a popular sight in arid gardens with their skinny trunks with a cluster of fronds at the top. However, unlike other plants, they have a specific time of year when they need to be planted. They require warm soil when planted, typically 65°F (18°C) or over, and late spring and early summer are ideal times for adding palms to your garden.

THE SEASONS IN THE DRY CLIMATE GARDEN

If you ask someone who lives in an arid or semi-arid region how many seasons they experience, you'll likely hear, "there are two seasons—hot and warm." On the surface, it's easy to see why that is a standard reply, especially if you originally come from a region with four distinct seasons heralded with colorful foliage, snow, and the rebirth of plants in spring. However, we do have seasons, just with more subtle differences. Ironically, some desert regions experience *five* different seasons.

Seasons aren't all the same length of time in terms of temperature averages. Hot temperatures generally arrive in mid to late spring and extend into autumn. We also enjoy a "second spring" when summer temperatures wane, and we venture out into the garden again. Winters can seem short, and sometimes it can seem like spring is skipped when hot summer temps arrive early.

Rain connects to the seasons in the garden, arriving in winter with isolated rain events in spring and fall. Unfortunately, most arid and semi-arid regions get little to no rain in summer, which is difficult for plants. Native plants usually do fine with this long dry stretch with no rain, but supplemental water is required when winter rains are absent. This rain-free period isn't usually a factor in how well plants are doing in your garden if they receive supplemental irrigation.

MONSOON IN THE DESERT
With the subtle seasonal changes and perception that there are two seasons in a dry climate, there are some desert regions that have five seasons—spring, dry summer, monsoon (wet) summer, fall, and winter. Monsoons are seasonal wind shifts that bring moisture from the sea to a few fortunate desert regions. The monsoon wind shift occurs in summer and brings rain and higher humidity, benefiting plants during the most stressful time.

High winds and torrential rain are hallmarks of monsoon storms, which occur over a short period—often less than an hour in duration. Frustratingly, summer storms can be very localized, with one neighborhood receiving welcome rainfall while a mile away remains dry. However, rain isn't the only welcome benefit that monsoon summer brings—the increased humidity provides welcome relief to plants. The Southwest region of the United States and the southern areas of the Sahara Desert are two desert regions that experience monsoon summers.

Localized thunderstorms are common in mid to late summer in the desert Southwest region of the United States.

DRY CLIMATE SOILS

Soil may not be a fascinating topic for you, but it is for plants! It is a big deal as it provides them food, a place to store water, absorb oxygen, and support the plant. While you may think of soil as dirt, it isn't. Soil provides a complex mixture of organic matter, microorganisms, minerals, water, and air to plants. In dry climates, the soil has certain characteristics due to low rainfall amounts that are important to know so you can see if you need to modify it for plants or leave it alone (in most cases, you don't need to do anything).

Two characteristics of soils in arid regions are that they are alkaline and are low in organic matter. The good news is that native and desert-adapted plants do well in these soil conditions because they have evolved within these constraints. The texture of dry climate soils varies and can range from finely textured clay soil to sandy soil with large particles depending on the soil makeup for your region.

SOIL TYPES

It's essential to identify what type of soil you have in your garden, as watering needs vary depending on the texture of your soil. Clay soils consist of tiny particles and can be described as heavy and have smooth texture when wetted and rubbed between your fingers. A moistened clump of clay will squeeze out in ribbons between your fingers when moistened. Clay soils have tiny spaces in-between the particles for water and oxygen. As a result, they are slow to absorb water and can take several days to dry out after watering. In addition, rainfall and irrigation don't permeate the soil quickly, so it takes a long time for water to reach the root zone of plants.

In comparison to the texture of clay soil, sandy soil is made up of larger particles and has ample spaces for oxygen and water. Therefore, when mixed with water and held in your hand, they feel gritty in texture and, when squeezed, will fall apart. Often referred to as "well-drained soil," water reaches the roots of plants quickly, and the soil dries out very fast. Therefore, plants in sandy soils need to be watered more frequently than those in clay soil.

Silt describes soil that has particles larger than clay and smaller than sand. It has a smooth feel when dry. When wet, silt creates slick mud and is easily moldable. Soil described as "loam" is made up of clay, sand, and silt.

Sand

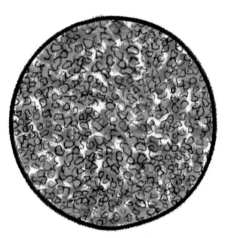

Silt

Clay

YOUR SOIL MAY HAVE COME FROM SOMEWHERE ELSE

The soil in your yard is likely not the native soil of your region. Agriculture and construction practices change the soil in a given area. When a home is built, topsoil is often removed and the remaining soil is compacted. "Fill dirt" is then brought in to level out the area. As a result, the soil texture in your garden can be different from what you expect.

The most crucial soil factor for most arid-adapted plants is well-drained soil. Plants need oxygen to survive, and they get it from the atmosphere and the soil. Oxygen is present in the tiny spaces between soil particles. Plants from dry regions have a high oxygen requirement, so they get a lot of that from the soil. The increased oxygen requirement is why you will often see "plant in well-drained soil" under the needs of most arid plant natives, as there is plenty of oxygen in-between those soil particles.

Soil testing for nutrients in your soil is rarely needed if you use native or desert-adapted plants, as these plants do well with low levels of nutrients unless they show signs of nutrient deficiency. Fertilizing plants at planting time isn't recommended as it stimulates growth before a plant has a good root system to support it. See Fertilizing section in Chapter 3 for fertilizer guidelines.

ALKALINE SOILS AND HOW THEY AFFECT PLANTS

Dry climate regions have alkaline soils, directly related to low rainfall amounts. Conversely, moderate to high rainfall areas usually have acidic soil because rainwater is slightly acidic. Most plants like to grow in somewhat neutral soil with a pH of 7, and soil is alkaline when the pH is over 7.5. The pH of your soil directly affects your plants because it affects the availability of nutrients in the ground—even if they are present in adequate amounts. Plants will struggle to survive and show signs of deficiency if they don't receive the necessary nutrients.

The good news is that if you use native or desert-adapted plants in your landscape, they will usually do just fine growing in alkaline soil. However, suppose you add plants that are native to regions with soils that are neutral or have slightly acidic pH. In that case, they can have a hard time growing in alkaline soil without amendments and regular fertilizing.

HOW TO CREATE THE PERFECT HOLE

You may not have given much thought to planting holes in the past. Most people dig a hole roughly the same size as the plant's root ball and do little else. But I'm here to tell you that the planting hole is a big deal and can significantly impact your plants and how quickly they grow. A well-prepared hole will allow a plant's roots to penetrate the soil easily, which is crucial for your plant's health and growth rate. For plant holes, it comes down to the size of the hole and whether to add amendments or not. In most cases, a shovel is all you need to dig a hole, but you may need to use a pickaxe or even a jackhammer to create a hole in very rocky soil.

THE RIGHT SIZED HOLE

Before we talk about measurements for your plant hole, it's important to discuss how a plant's roots grow. Most of the roots extend outward within the top 1 to 2 feet (30–61 cm) of soil, where most oxygen and nutrients are present. Larger plants, such as trees, will have roots that grow deeper to help anchor the plant. So, it makes sense that we should create wide holes for plants as roots have an easier time growing into loosened soil than hardpacked soil. A general guideline is to measure how wide the root ball of your plant is and create a hole that is three times wider. For example, a plant in a 1-gallon (4L) container is about 6 inches (15 cm) (wide, so the hole should be at least 18 inches (46 cm) wide.

While the width of the hole should be wider than the root ball, the same is not true for the depth of the hole. The hole should be as deep as the root ball or even an inch (3 cm) shallower to allow for settling after planting. This planting

The ideal hole should be three times wider than the root ball to allow plants to quickly grow a healthy root system. This is especially vital in dry climate regions where root growth is crucial to surviving the summer heat.

depth is because the plants' oxygen and nutrients are plentiful in the top several inches of soil. As you move deeper, less oxygen is present, which can cause stress for plants that are planted at too great a depth—even a few inches can make a difference.

A common mistake is when trees are planted too deeply. Sometimes, a few inches of excess soil or mulch cover the tree's base in its nursery container. Before planting, using a hand shovel, gently scrape away the excess dirt from around your tree until you see the trunk begin to flare slightly outward at the base or look for the presence of tiny root hairs. Once you see one of these signs, you will know the correct depth for planting your tree.

As discussed earlier in this chapter, well-drained soil is vital for plants adapted to dry climates.

So, before planting, we need to make sure that our soil drains fast enough for these plants to grow. To determine if your soil is well-drained or heavy, dig a hole 12 to 18 inches (30–46 cm) deep and wide. Then, fill with water in the morning, return later, and fill again. Twenty-four hours after filling the hole a second time, all the water should be gone—if it isn't, plants are likely to struggle, and strategies or amendments can improve drainage before adding plants.

MANAGING CALICHE IN SOIL

There are many unique factors when gardening in a dry climate, and one is the presence of caliche in the soil. Caliche is a cementlike layer in the ground that can make growing plants difficult. You've likely hit caliche if you have ever dug a hole and run into a rocklike layer. This hard layer forms when calcium carbonate, already present in the soil, binds clay or sand particles into a rock-hard layer that is hard for water and roots to penetrate. There is variability in this layer, which can be a couple of inches

(or centimeters) thin or much thicker. Caliche can occur at the surface, several inches below, or a few feet underneath the soil. It can cover a large or small portion of your yard. In many cases, you don't know you have it until you try to dig a hole and run into an impervious layer of rock.

Caliche can make it extremely difficult, if not impossible, to grow plants above it because water doesn't drain through it. This lack of drainage leads to wet soils and a lack of oxygen for plants, and since most arid-adapted plants require well-drained soil, they struggle to survive. So, what can you do if you find caliche in your soil? Is it possible to grow anything? The answer is "yes" if the layer is just a few inches (or centimeters) thick, and you can break through it. However, if it is thicker than that, it will be difficult to grow anything in that area, and you may want to add potted plants in that area instead.

How to maximize success when adding new plants in areas with a caliche layer.

Soil

Caliche

The key to growing plants where caliche is present is to create small drainage holes through the caliche layer that allows water to drain and prevents waterlogged soil. This only works if the caliche layer is thin enough to punch a hole through using a pickaxe or heavy iron bar. If the caliche layer is only 1 or 2 feet (30–61 cm) under the soil, you will want to grow shallow-rooted plants like shrubs, groundcovers, and succulents. You can also create a mounded soil area over caliche areas to provide more room for root growth and drainage. The hole for your plants should be three times larger than for areas without caliche, so plants have ample room to grow their roots and compensate for the limitations that caliche presents to root growth. After you dig the hole, fill it with water and watch to see that the water level sinks at least 1 inch (3 cm) an hour. This test will ensure that there will be enough drainage for your plants without breaking through the caliche. Before planting, amend the soil with a one-part compost ratio mixed with two-parts existing soil, which will help improve drainage.

TIPS FOR SUCCESSFUL TRANSPLANTING
Water your plant deeply a day or two before moving it to allow it to get a big drink of water to shore up its reserves and make the soil easier to penetrate with your shovel. At the same time, dig the new hole and fill it with water. Adding water to the hole allows you to test for adequate drainage and will ensure a reserve of soil moisture for the new plant. When it's time to dig up the plant, take care to remove as much of the roots as possible with the soil intact, so, for a shrub that is 2 feet (61 cm) wide, make sure the root ball is 2 feet (61 cm) wide. Gently place the plant in its new hole and try

not to jostle the roots. Fill in the rest of the hole with existing soil, tamping it down to remove any air pockets. Water deeply with your hose on a slow trickle.

When you transplant groundcovers, perennials, or shrubs, wilting and some leaf drop is expected because the reduced size of the root system can't support the original size of the plant. If this happens, you will want to prune off some of the foliage to reduce the amount of water stress. A general guideline is to prune up to half of the outer growth. Occasionally, plants will drop all their leaves but grow new ones to replace them. It can take three months or more before a plant will leaf out again. Sometimes, plants won't survive the process despite doing all the right things to maximize transplant success. If that happens, plant a new one from the nursery.

If the new spot for your plant receives more sunlight than the old one, you may need to provide temporary shade protection using a shade cloth through the first summer—particularly if you plant in early spring. However, it is best to wait until late fall, late winter, or early spring for the best results. In other words, don't transplant when it is warm or hot outdoors, as the higher temperatures will increase the intensity of water stress for the plant and make it less likely to survive.

PLANTING TIPS FOR CACTI AND OTHER SUCCULENTS
While planting trees and shrubs is pretty straightforward, adding cacti and other succulents can present some unique challenges, and specialized strategies come in handy, mainly when dealing with prickly spines.

PROPAGATING CACTUS

Many types of cacti, like prickly pear species (*Opuntia spp.*), and columnar cacti, such as Mexican fence post (*Pachycereus marginatus*), can be started from cuttings when temperatures are over 60°F (16°C)—ideally in late summer. Cut off a pad from a prickly pear cactus or a long stem from a columnar cactus with a sharp knife. Place the cutting in a dry shady spot for two weeks to allow the cut end to dry entirely before planting it in its new location. Be sure to point the cactus cutting in the same direction as previously. Wait a month before watering.

Spring and early summer are ideal for adding succulents (including cacti), but you can also plant them in late summer. This timing ensures that soil temperatures are warm enough to foster new root growth, vital for success. Don't plant them in winter as it is too cold for new root growth, and there is an increased chance that the succulent will die. All succulents need well-drained soil so if you have heavy clay soil, amend the soil in the hole with one part of compost to three parts of existing soil to help improve soil drainage.

When you plant a cactus, you want to orient the cactus in the same direction it was initially. The reason is that the sides of the cacti that face south and west adapt to handle the intensity of the sun's rays, while the east- and north-facing parts will experience sunburn if put in full sun exposure. Most cacti nurseries will mark the orientation of cacti on the nursery container with a dot of paint or chalk mark, so you know what direction to situate them in their new spot. However, suppose you don't know its original exposure. You can cover the cactus with a 30 percent shade cloth in its first summer and then remove it in fall to allow the cactus to gradually become acclimated to its new exposure when the sun's rays are less intense.

So, how do you plant a cactus without getting hurt? Carpet remnants, newspaper, rubber straps, tongs, and an old towel are household items that are frequently used to plant cactus. Relying on a pair of gloves to protect you from the prickly spines of a cactus usually isn't enough. Direct contact with a cactus will result with spines embedded in your gloves, ruining them. An old set of kitchen tongs or a towel (folded into quarters) is what I use to plant potted cacti.

A piece of carpet or rubber straps are useful when maneuvering a larger cactus into place.

STAKING TREES

Young trees may require staking when first planted in your garden, especially if you bought one in a 24-inch (61 cm) box or larger. Staking is needed when trees don't have enough roots to support the top growth, or their trunk may be too narrow. Trees purchased in a smaller-sized container rarely need stakes and tend to grow faster than older ones in larger-sized nursery containers. Other trees that may not require staking include multi-trunk tree specimens.

If you need support for your tree, it's important to remember that staking is a *temporary* support strategy while a tree grows a robust root system and trunk strength to support itself. The easiest method consists of using two wood or metal stakes. Insert one stake on either side of the tree about 3 feet (91 cm) from the trunk. Orient the stakes so that they are perpendicular to the wind direction. Tie the tree to the stakes, halfway to two-thirds up the height of the trunk. Wire is a good choice for tying the tree as long as you use a section of hose to cover the wire that encircles the tree to protect the trunk. If your tree came with a stake tied up against its trunk, remove it as it will impede the tree's future growth. Instead, allow for some tree movement, building strength in the trunk. Check the ties every few months and loosen them as needed to avoid cutting into the trunk. Untie the tree after two years (or earlier) and see if it is stable. If so, then remove the stakes. A tree should rarely need staking after two years unless it has a weak root system.

A newly planted young blue palo verde (*Parkinsonia florida*) with tree stakes for temporary support.

WHEN TO ADD SOIL AMENDMENTS AND WHEN NOT TO

A common practice is to add amendments in the planting hole to create an ideal environment for the roots of your new plant. Amendments are products added to the soil, such as compost, manure, and fertilizer. However, most studies have shown that using amendments in a plant hole can impede the growth of your new plants. The reason is that amendments create an ideal place for a plant's roots, and they don't want to grow beyond that. A healthy plant needs a robust root system that extends beyond the plant and not one that

remains localized to the original plant hole. No amendments should be added for planting unless growing annual flowers, roses, or vegetables. There are, however, some instances where compost, an organic amendment, should be added to a planting hole. These include helping to improve poorly drained clay soil or very fast-draining sandy soil. Mix one part of compost with two parts of existing soil in the hole. The compost will help break up the heaviness of clay soils and allow them to drain better. With sandy soils, compost helps it to retain water. Be careful not to only plant in compost alone—mix it with the existing soil to allow for outward root growth. Compost is readily available at most nurseries and garden centers if you don't make it yourself.

Contrary to widespread practice, you should not add fertilizer to holes for new plants. When adding a new plant to your garden, it needs to grow its roots first. Adding fertilizer when you plant stimulates the growth of branches, stems, and leaves when the plant's root system may not be large enough to support it. For native and desert-adapted plants, most don't need supplemental fertilizer anyway.

New plants will require more water initially after planting to help them to grow their roots, which is needed for them to become considered established. After planting, tamp down the soil around the plant to remove any air pockets and then provide them with a deep drink of water. Place a hose on a slow trickle by the plant and thoroughly wet the entire root zone. Be sure to connect to your irrigation system (if you have one) or follow up with hand-watering—see Watering New Plants section in this chapter for more guidelines on watering new plants.

The author plants vinca flowering annuals in soil amended heavily with compost.

WATERING NEEDS FOR NEW PLANTS

New plants need extra water to grow a sound root system that will allow them to exist on less water in the future, which applies to native plants as well. If your new plant does not connect to an irrigation system, create a basin around it that you can fill up with water and help hold the water around the roots. Initially, plants may need daily watering for two to three weeks, which can slowly decrease in frequency over the next few months until you reach recommended timing for your region.

For newly planted trees, create a basin that extends to the edges of the branches. The sides of the basin should be 3 inches (8 cm) high. After planting, fill the basin and fill it again later in the day. Water daily for the first week, then reduce the watering frequency to twice a week for the next couple of months. Keep an eye out for yellowing foliage or the unusual dropping of leaves, which is a sign of overwatering. The goal is to gradually move toward the irrigation schedule tailored for your area within two to three years. You can usually find watering schedules online through your water provider.

People make a common mistake in that they never decrease how many times they water their plants—instead, they keep it on the new plant watering schedule. If you give your plant more water than they need, they can slowly drown due to a lack of oxygen around the roots.

Use what you already have on hand to provide temporary shade for plants recently transplanted. Patio umbrellas offer excellent shade during a heat wave.

MULCHES

In arid and semi-arid climates, mulching is a beneficial practice to help minimize the stress of hot, dry summers on plants. Mulch is added on top of the soil around plants to assist plants in various ways. The most important is that they slow down the evaporation of water around the root zone of plants by protecting them from the sun's rays. They also help moderate soil temperatures by keeping them cooler in summer and warmer in winter. Mulch also decreases the number of weeds that come up. Many different materials make a suitable mulch, but first, let's start with the two main types of mulch—organic and inorganic.

DON'T THROW OUT YOUR LEAVES

A typical maintenance task involves the use of leaf blowers, which clean the landscape by blowing leaves away. However, these leaves are nature's way of replenishing the soil with nutrients and creating a natural mulch. So, instead of removing all fallen leaves, rake (or blow) them around the base of plants where they will gradually break down, enriching the soil—free mulch!

ORGANIC MULCH

This type of mulch comes from living materials such as compost, bark, grass, leaves, and straw. After applying, it slowly breaks down into the soil, adding nutrients and improving soil texture over time. However, the larger the size of the material used, the longer it will take it to break down, in the case of bark. Therefore, fine-textured organic mulch needs to be refreshed often as it breaks down and you begin to see the soil surface. An excellent benefit of organic mulches is that they enrich the soil over time, improving the texture and adding nutrients.

Bark mulch is used throughout this area surrounding the striking artichoke agave (*Agave parryi v. truncata*).

A desert garden with various sizes of rock mulch, including ½-inch (1 cm) landscape rock, riprap, and boulders, which add texture and a design element.

areas where fire danger is prevalent, landscape rock is a favored choice as it is fire-resistant. As opposed to organic mulches, inorganic mulches last much longer but do not enrich the soil. Landscape rocks decrease soil temperatures and help slow water evaporation. Yet they increase the temperature above ground by several degrees as they absorb the heat from the sun and re-radiate it toward plants. This heating benefits frost-tender plants in winter but can stress ill-adapted plants in summer. Shade from trees, planting groups of shrubs, and groundcovers can help mediate the effects of the heat from inorganic mulches.

When adding new plants to a garden with rock mulch, expose the soil's surface by raking back the rock a few feet wider than the hole to allow room for the pile of dirt you dig up. Then, after planting, you can rake the rock mulch to the base of the plant, or you can add an organic mulch around the plant and leave the rock mulch farther out. Having localized areas of organic mulch around plants in a landscape with rock mulch is a useful option if you have a plant that is sensitive to reflected heat or needs more fertile soil.

Apply 2–3 inches (5–8 cm) of mulch around plants in the spring. Mulch should cover the root zone of the plants, extending out to the entire width of the plant. Take care to keep the mulch a few inches away from the main trunk of trees to avoid creating an ideal environment for harmful fungi or insects.

INORGANIC MULCH

Drive down any desert neighborhood, and you are likely to see landscapes covered in a layer of rock, which is a form of inorganic mulch. Commonly called "gravel" or "landscape rock," it is the most popular mulch in the desert and is increasingly popular in semi-arid regions. In

Landscape rock is available in different colors and sizes. Thankfully, gone are the days when the blue and green stones were in fashion. Instead, rock in earth-tone shades ranging from tan to brown is preferred. The size of landscape rock ranges from ¼-inch (6 mm) in size to 9 inches (23 cm). When ordering, the size refers to the average size of each little rock, with some slightly smaller and larger than the median size. The smaller the stone, the more area it covers versus larger rock sizes.

Decomposed granite covers the author's backyard landscape, creating a pleasing backdrop for her colorful plants.

With so many size choices available, how do you decide which is best for you? Here is a breakdown of the most popular types beginning from smallest to largest:

Decomposed Granite (DG)

Decomposed granite, referred to as "DG," is ¼-inch (6 mm) in size and is often used in landscapes with native plants as it mimics the ground in nearby natural areas. This small-sized landscape rock is also a favorite choice for creating walkable pathways and can be mixed with a stabilizer to make a hard surface. DG is also utilized for patios and driveways and works well in desert and Mediterranean-style landscapes. Avoid decomposed granite if you have wood floors, as it will scratch them if you track it in on your shoes. If you have areas in your landscape that get washed out when it rains, a larger rock would be better as it won't move as easily.

½-Inch (1 cm) Landscape Rock

Where I live, this is the most popular size of rock used by homeowners. It is relatively easy to walk on (with shoes), provides good coverage over the soil, and doesn't wash away quickly when added to mostly level landscapes. Another plus is that it isn't as dusty as DG and doesn't track easily into the house.

¾- to 1-Inch (2–3 cm) Landscape Rock

Landscape rock that ¾-to 1-inch (2–3 cm) is an excellent selection for those who want their rock to be more of a landscape feature instead of staying in the background. However, getting around the garden is slightly more difficult as the surface is more uneven than walking over smaller rocks.

Over time, landscape rock will need to be refreshed as dirt will cover the lower layers of the rock. You will know when to add more stone when you see glimpses of soil through the rock. When ordering, you will need to get enough to cover at a 1-inch (3 cm) depth since there is still existing rock matching the same color and size of stone as before. If you decide to change your rock, you'll need to remove all the old rock and add new. Otherwise, the old and new rock will mix and create a messy appearance if they are very different in color and size.

3- to 9-Inch (8–23 cm) Riprap or River Rock

Riprap is typically an angular granite, while river rock has smoothly rounded edges. Larger rock sizes are a common feature for areas subjected to periodic runoff from rainfall as they don't wash away easily. Large rock is often used as a design element in areas like a swale or a curved dry creek bed that winds through the garden. Over time, the bigger rock will accumulate

WHY PLASTIC LANDSCAPE FABRIC IS A BAD IDEA

Years ago, using sheets of plastic laid over the soil before applying landscape rock was a common practice. Sadly, even though it isn't as popular, it's still done. The purpose of the plastic is to prevent weeds from growing through the rock. However, it is only marginally effective at weed prevention and has many adverse effects on plants and the soil, which include the following:

- It prevents rain from penetrating the soil and decreases soil fertility.
- Oxygen in the soil is depleted, leading to plant stress.
- Living organisms in the soil that benefit plants struggle to survive.
- The soil becomes sterile and not conducive to growing plants.
- Plant root growth is restricted due to a lack of water underneath the plastic sheeting.
- Over time, the plastic becomes torn and comes up in unsightly pieces.

While it will keep weeds away for a short while, plant debris and dust will filter down through the rock to rest on top of the plastic, where weeds will eventually take root.

To sum it up, skip the plastic.

River rock surrounds young 'Blue Glow' agave (Agave 'Blue Glow'), creating a decorative filler in a narrow planting area in a front entry.

dirt, smaller stones, and weeds in between the individual stones and will need to be cleaned out. The cleaning process can be tedious, but it is simple and involves removing each rock, cleaning it off with water, re-grading the area (if needed), and setting the stones back.

One last option you will need to consider is whether you purchase a rock product that includes finely crushed rock (gravel) intermixed with the larger pieces. Usually called "fines" or "minus," they are available in many ¼-inch (6 mm) to 1-inch (3 cm) landscape rock mixes. The fines filter down through the larger rock to help form a base underneath that helps keep the bigger stone more stable and less likely to come loose onto the sidewalk or move. Screened gravel doesn't include many fines, but the rock will be more open and move more when walked on or scattered onto your patio or walkway. I have used both types and tend to favor the landscape rock that includes fines so it doesn't move too freely, but both are suitable for dry climate landscapes.

Whatever size rock you select, you will need to figure out how much you will need. To determine this, you will need to measure the square footage of the area to be covered. Using the chart to the right, select your rock type and the area a ton will cover.

HERE ARE THE FORMULAS YOU WILL NEED:

Length × width = square footage
Square footage / Area covered per ton = tons of rock needed to cover the area

EXAMPLE: Using the chart below as a reference—an area that measures 100 feet long (30 m) and 60 feet (18 m) wide needs ½-inch (1 cm) landscape rock at the recommended 2-inch (5 cm) depth.

Area: 100 (30 m) × 60 (18 m) = 6,000 square feet (30 × 18 = 540 square meters)

Amount covered per ton: 120 square feet (11 sq. m)

Tons of gravel needed: 6,000 / 120 = 50 tons (45 metric tons) of ½-inch (1) landscape rock needed for coverage.

ROCK TYPE	RECOMMENDED DEPTH	AREA COVERED PER TON OF ROCK
3"–9" (8–23 cm) riprap rock or river rock	3"–9" (8–23 cm)	60 sq. feet (6 sq. m)
1"–3" (3–8 cm) landscape rock	1"–3" (3–8 cm)	80 sq. feet (7 sq. m)
¾"–1" (2–3 cm) landscape rock	2" (5 cm)	100 sq. feet (9 sq. m)
½"–⅝" rock (1–2 cm)	2" (5 cm)	120 sq. feet (11 sq. m)
¼" (6 mm) – decomposed granite (DG)	2" (5 cm)	140 sq. feet (13 sq. m)

The color of the rock you use is a matter of personal preference. I recommend using stone in beige or light tan shades as I prefer plants to be the focus in my garden and not the rock. While darker colors, such as deep browns, are attractive, they get hotter than rock in lighter shades. If you live in a neighborhood with a homeowners' association, check to see if they have requirements for the size and color of landscape rock you can use to avoid a costly mistake.

CHAPTER 3

CARING FOR PLANTS IN THE DRY CLIMATE GARDEN

I often hear people complain about how much work they need to do in their garden. In other words, they want an attractive landscape, but they don't want to spend lots of time and money to have a beautiful outdoor space. Is it possible to have a low-maintenance, water-wise landscape that looks great? Yes! That is one of the many benefits of having a garden populated with plants native to arid regions. In addition, native plants have particular adaptations to help them grow in dry conditions, so they don't need much care if appropriately chosen.

Surprisingly, most people do more for their plants than they need. In other words, you may be overcomplicating the upkeep of your garden by watering too often, pruning several times a year, and fertilizing plants that don't need it. As a result, the over-maintaining of plants can lessen your enjoyment of your outdoor space.

There are two main components to having a beautiful garden without excessive maintenance. The first step is selecting the right plants for the climate. The second is learning to care for your plants correctly. For example, pruning and watering are the primary maintenance tasks in the dry climate garden, yet most people don't know the right way to prune or water. Therefore, it's essential to understand how to do them correctly. Other garden responsibilities can include cold protection and pest control, which aren't as crucial if you select native plants or those resistant to common pests.

Low-maintenance beauty is achieved in the landscape using desert milkweed (*Asclepias subulata*), desert marigold (*Baileya multiradiata*), Sierra Star® fairy duster (*Calliandra* × Sierra Star®), and brittlebush (*Encelia farinosa*), with deer grass (*Muhlenbergia rigens*) in the background.

An attractive landscape with pink fairy duster (*Calliandra eriophylla*) and feathery cassia (*Senna artemisioides*) adds late winter color to the desert garden and requires very little maintenance.

A plant's requirements are simple—water, sun, and food. Note the natural terrain in your region and how well the plants do with little to no help. However, the truth is that we often complicate things within our gardens by over-maintaining our plants. As a horticulturist and landscape consultant, I meet countless clients who struggle with their plants because they do too much with little to show for their efforts. Our human nature makes us want to nurture our plants even when they don't need it. The results are over-pruned shrubs, plants with lank growth due to overfertilization, yellowing leaves from overwatering, and a frustrated homeowner.

One of the most rewarding parts of my work is showing people the "right" ways to care for their plants, which usually involves pruning less often, ditching the fertilizer, and watering less frequently. Imagine their shock when they see their plants begin to thrive and look better with less effort! In this chapter, I will share with you how to properly maintain your plants because they do need some care from you from time to time to help them look their best.

A front yard filled with flowering shrubs that have been incorrectly pruned into unnatural shapes, stripping them of their beauty.

PRUNING

Trees, shrubs, and other leafy plants need pruning to keep them healthy and structurally sound. They don't need to be pruned into unnatural shapes that turn them into green "blobs" dotting your garden. First, this isn't healthy for the plants themselves; it is ugly and wastes unneeded time and money. Sadly, there is an epidemic of this type of pruning practice. So much so that it has become accepted as a standard way to maintain plants. In the industry, the tendency for people (including landscape workers) to shear shrubs (and sometimes trees) into the shape of balls, cupcakes, and squares is called "poodle pruning."

We add plants outside our homes to add beauty, but frequent, formal pruning does just the opposite as it strips the shrubs of their unique shapes that add welcome texture to the landscape. It also removes the buds of flowering shrubs before they begin to open. Does this sound like you? Let's discuss why formal pruning isn't recommended for flowering shrubs.

WHY PRUNING TOO OFTEN IS BAD FOR SHRUBS

The leaves of plants aren't just for show; they are the part of the plant that makes the food for the plant. A shrub that is allowed to grow to its natural shape and size provides shade to the root zone, moderates soil temperatures, and helps to slow down the evaporation of water from the soil. When you prune a shrub, new growth soon follows, producing fresh leaves to continue to make food for itself. With infrequent pruning, this works well. However, when excessive pruning, as soon as the shrub grows back some new leaves, they are cut off, so it grows new ones so it can survive only to be removed again—and the cycle repeats.

With this maltreatment, it is no surprise that plants maintained like this have a shorter lifespan than those pruned less often.

Frequent pruning makes shrubs grow faster as they constantly try to replace their leaves. In addition, they also require more water because they are actively growing all the time. Finally, let's be honest. Would you rather have a front yard filled with green blobs or one with beautiful shape, color and texture with a fraction of the effort?

THE RIGHT REASONS TO PRUNE

Pruning plants isn't harmful when done correctly and for the right reasons. The primary goal of

A tale of two 'Green Cloud' Texas sage shrubs (*Leucophyllum frutescens* 'Green Cloud'). The one on the left is pruned into a green ball, and five minutes later, the one on the right ends up the same, removing its flowers and natural shape.

Repeated shearing of shrubs creates a virtually impenetrable barrier so sunlight cannot reach the interior. As a result, a buildup of dead areas accumulates that enlarges over time, causing unsightly dead spots and the eventual early death of shrubs.

pruning trees and shrubs is to remove dead or diseased growth. It can also help train a plant to grow upward against a wall or gradually train a large shrub into a tree shape. Pruning removes old, woody branches that are unproductive or weakly attached to tree trunks. In all cases, avoid pruning to keep a plant confined into too small a space if constant pruning is required. When approaching pruning your plants, keep in mind that the ultimate goal of pruning is to encourage the plant's natural shape while keeping it healthy.

HOW TO HELP YOUR GARDENER PRUNE YOUR PLANTS THE RIGHT WAY
When you hire a gardener or landscape professional to maintain your garden, you assume they know the right way to prune. Sadly, this is often not the case. I have seen countless plants turned into anonymous green blobs by landscape workers.

So, what can you do to ensure that you prune your trees and shrubs the right way? First, find a landscape professional who knows the right way to prune. One way to do this is to look at the front yards in your neighborhood and identify ones with healthy trees and shrubs growing into natural shapes. Then ask your neighbor who their gardener is and hire them to care for your plants.

Secondly, suppose you have difficulty finding a knowledgeable gardener. In that case, you can work with the one you currently have, which involves getting involved with the upkeep of your landscape instead of blindly trusting them to know the proper maintenance practices intuitively. Instruct them not to prune any plants unless you tell them. Tell your landscaper what plants you want to be pruned and how you want it done when it is time to prune. It may be that you prune a single shrub yourself to show them how you want them to do the rest. Or, direct them while they are pruning your shrubs. You can tell them how far back you want your shrubs pruned, which is usually sufficient for them to understand what you want.

Frost-damaged bougainvillea shrub

Pruning back shrub using loppers in spring

After severe pruning

Bougainvillea shrub eight months later

Finally, you can forgo having a gardener and prune your plants yourself. Many homeowners I work with like to do this task themselves once they know the correct methods, as they get to spend time outdoors, get some exercise, and feel a sense of accomplishment when their efforts result in a beautiful outdoor space filled with healthy plants.

THE BEST TIMES OF YEAR TO PRUNE

The timing of pruning is essential as we want to do it when the plant is ready to grow. There are less than ideal times of year to prune, and if you prune at the wrong time, plants will struggle or even die because of it. Not all plants have the same schedule when it comes to pruning—it varies depending on the type of plant and their specific needs. However, it is possible to break pruning seasons up into broad categories to enable you to know the best time to break out your pruning tools.

When to Prune Trees

To determine when to prune trees, we need to separate trees into three separate groups. The first group is evergreen to semi-evergreen trees, meaning that they retain most of their leaves through the year—even in winter. For

trees that match this description, pruning is best done in spring once the threat of freezing temperatures has passed. Sometimes, additional pruning may need to be done in late summer if overgrowth becomes evident—this can be particularly true of younger trees.

Deciduous trees lose their foliage in winter and should be pruned in late winter to early spring, just before their leaf buds swell. At this point, they will still be leafless, which makes it easier to see what to prune to promote a pleasing shape and sound structure for the branches.

If you live in an area where summer storms or high winds are typical, it is essential to have your trees pruned before the windy season arrives. The focus is to identify weakly attached branches and remove any excessive growth that may lead to large branches breaking off.

It is important to note that not all trees need pruning every year, which is particularly true for old trees, which aren't growing as actively as younger ones. I advise you to consult with a certified arborist who receives specialized training to help you determine when and if your tree needs pruning.

Valentine bush (*Eremophila maculata* 'Valentine') and yellow-blooming feathery cassia (*Senna artemisoides*) bloom once a year in spring and are pruned once their flowers fade. Trailing lantana (*Lantana montevidensis*) blooms throughout the year and is pruned in spring.

Although they are not technically a tree, palm trees are a common sight in many dry climate gardens, where they are used as accent plants for interest. The window for pruning old palm fronds is late summer and early fall.

The Best Time(s) to Prune Groundcovers, Shrubs, Vines, and Ornamental Grasses

Spring is the busiest time for pruning in the dry climate garden. The reason is that temperatures are beginning to rise, and plants are ready to put on a flush of new growth and generally respond well to pruning this time of year. It is crucial to wait until the last freezing temperature has passed for your area before pruning frost-tender plants, as a late freeze can damage the new growth of your plants.

We need to break up spring into two parts for pruning—early spring and late spring/early summer. Plants that bloom in spring, summer, and perhaps into fall can be pruned in early spring once freezing temperatures have passed. Occasionally, they may need some light pruning from late summer to early fall if they begin to outgrow their allotted space. However, finish all pruning at least eight weeks before your region's first average frost date. Avoid pruning in summer as the intense heat and low humidity can severely stress plants. Winter pruning isn't necessary except for specialty plants such as roses.

A collection of trees such as desert willow (*Chilopsis linearis*), ironwood (*Olneya tesota*), foothills palo verde (*Parkinsonia microphylla*), and native mesquite (*Prosopis velutina*) add ornamental beauty and welcome shade to this desert garden.

The next group of plants flower once a year, in late winter or spring, and then cease midsummer. We want to prune these plants only once a year when they have stopped blooming. They don't need any other pruning throughout the year. Doing so will remove the branches that will produce next year's flower buds.

Ornamental grasses are a popular feature of arid landscapes with their flowing shapes and finely textured foliage. These are a great low-maintenance choice as they only need pruning once a year in early spring.

SKIP YEARLY PRUNING?

Not all shrubs, groundcovers, or vines require annual pruning. You can skip pruning for a year or more if they have enough room to grow and don't have a buildup of old, woody growth. Wait until you have a reason to prune rather than default to automatic yearly pruning. You will save time and effort and your plants will look better!

PRUNING TOOLS YOU SHOULD HAVE

The proper pruning tools are crucial to pruning plants the right way. Using the wrong tools can injure your plants and make pruning more difficult for you. Here are the five most-used pruning tools for dry climate gardens:

- HAND PRUNERS are used for cutting stems and branches up to a ½-inch (1 cm) thick. Useful for deadheading flowers, removing small branches, and cutting flowers to bring indoors.
- LOPPERS have long handles and cut branches up to 1 inch (3 cm) in diameter. They are used to prune young tree branches, cut back larger shrub branches, and remove palm tree fronds from shorter palm tree species.
- PRUNING SAWS are helpful in removing medium-sized tree branches and thick shrub branches that are too thick for loppers.

- TREE POLE PRUNER has a telescoping handle that allows you to reach taller branches of trees from lower down and a pruning saw for larger branches.
- HEDGE TRIMMERS are often misused to poodle prune shrubs into unnatural shapes, but they can be used to prune plants correctly. Use to prune grasses, groundcovers, and small shrubs back severely in spring.

There are pruning tool(s) for the maintenance needs of your plants. Remember, there aren't bad pruning tools, just people who use them incorrectly.

The author removing a damaged branch from a desert willow (*Chilopsis linearis*) tree.

HOW TO PRUNE PLANTS THE RIGHT WAY

Let's talk about the correct ways to prune plants. These recommendations are likely different from what you have done in the past and are much easier than you think, involving less time and effort, resulting in healthy and beautiful plants.

PRUNING TREES

Trees are the cornerstone of our landscape, which is why a badly pruned tree stands out like a sore thumb. Unlike other plants, they can struggle to overcome the damage caused by being pruned the wrong way. As a retired certified arborist, I have seen trees disfigured and inadvertently killed by well-intentioned homeowners and landscapers. Sadly, this is due to the homeowner not understanding how and why trees should be pruned or trusting someone else's expertise who lacks the knowledge of how to prune correctly.

Let's break down tree pruning to help avoid unfortunate pruning mishaps. First, keep in mind that the ultimate goal of tree pruning is to remove dead, diseased, or crossing branches. Standard trees have a single trunk and are more susceptible to wind damage. Many trees are available with multiple trunks, which helps distribute the branches' weight, making them less prone to wind damage.

The amount of foliage you remove counts when you prune trees. As discussed earlier, leaves are how a tree makes food. If we remove too many, it can stress the tree, making it susceptible to insect damage, sunburn, and weak wood formation. The standard guideline is to remove no more than 20 percent of a tree's growth annually.

Tree Suckers

Many trees from arid regions require pruning to create and maintain a traditional tree shape, which involves removing "suckers," which are small branches that grow out from the base of the trunk. Suckers are removed with hand pruners and can be pruned anytime in the growing season. For younger trees, allow the suckers to remain for the first few years as they help promote the tree's overall growth and contribute to its trunk strength. Approximately three years after planting, gradually remove the smaller, lower branches in spring to help elevate the tree's canopy.

The author prunes away suckers from her desert willow tree.

Yearly Tree Pruning

Remove any dead branches where the wood easily snaps when broken off, which signifies that part of the tree is dead. Next, look for any signs of disease within the branches and remove the branch at least 1 foot (30 cm) below where the disease occurs. Then step back to look for branches that cross or rub against another branch—remove one of the branches, keeping the healthier one that grows outward and not toward the tree's center.

Identify any long-reaching branches that may have a likelihood of breaking off in windy conditions. You can prune these overreaching limbs by pruning them back, making the cut right in front of another branch further back on the branch. Stand back from time to time as you are pruning to get an overall picture of what may still need pruning, focusing on removing 20 percent or less.

Trees add beauty to the outside of our homes and increase their value as buyers look for houses with mature trees. However, it can be intimidating or impossible to maintain trees yourselves due to their size or condition. I strongly recommend consulting with certified arborists trained in proper tree care, including pruning. The International Society of Arborists has excellent online information to help you care further for your trees and a list of certified arborists specific to where you live.

Common Tree Pruning Mistakes to Avoid

Never "top" a tree. This injurious pruning practice involves shearing off the top of the tree's branches. When a tree is topped, it leaves it open to sunburn and insect damage, and the branches that grow back will be weakly attached and in danger of breaking off. In addition, this type of pruning makes the tree grow back faster.

Don't use pruning paint to cover pruning cuts, which is an outdated practice, and recent studies have shown that it slows down healing. Trees do an excellent job of healing themselves and heal faster without pruning paint.

Never leave "stubs" behind when pruning. A stub occurs when a tree branch has been cut haphazardly. Unless there is another smaller branch next to the cut to take over and continue growing, the stub often browns and dies, creating a haven for damaging insects or disease.

Avoid over-pruning your trees. A common pruning mistake is "lion tailing," which occurs when too much interior growth has been pruned away from the tree. This results in bare branches with smaller clumps of foliage or tufts hanging onto the tips of branches. Sadly, this type of pruning mistake is prevalent because people like to see the branches and think it will help the wind flow through the tree instead of breaking the branches. However, it has the opposite effect. When too many smaller branches are removed from larger ones, it depletes the ability of the tree to make food for itself resulting in weak wood with branches more apt to break under windy conditions.

A badly pruned Chilean mesquite (*Prosopis chilensis*) tree. Too much foliage has been removed due to "lion tailing," and the remaining stubs are likely to attract disease and damaging insects.

WHY EXPERT TREE PRUNING IS THE BEST

Beware of trusting your tree's pruning needs to your gardener or landscaper. While they have good intentions, they don't always know the right way to prune. A tree that has been badly pruned can take years to recover, if at all. Suppose you see stubs on a tree after it has been pruned or one that has had more than 20 percent of its foliage removed; that is a clear sign that the person who pruned your tree doesn't know the right way.

The harmful practice of topping trees is formally pruning or shearing trees around their sides. Trees respond similarly as shrubs do to this pruning, and it makes them more prone to blowing over in high winds, and the winds cannot permeate the tree to blow through. Always prune back to another branch further up the limb when reducing the width of your trees. If your tree has been maintained incorrectly like this, engage the services of a certified arborist to rehabilitate maltreated trees like this.

Palm Tree Pruning

As a palm tree's fronds (leaves) age, they slowly droop while they fade to brown, and if left alone, they will eventually fall off. For California and Mexican palms, the dead fronds are retained for several years, and they form a "skirt" of brown palm tree leaves under the green fronds, which creates an iconic palm tree appearance.

There is nothing wrong with leaving the "skirt" on these types of palms, and it is a matter of preference whether to remove them. For other species of palm trees, I recommend removing the palm fronds once they have turned completely brown. At the same time, it may be tempting to remove those that still have some green left on them, but don't, as they are still producing food for the tree.

Prune palm trees in late summer using a pruning saw, serrated utility knife, or loppers for smaller fronds. You can also remove any seed-bearing stems at the same time. Avoid over-pruning your palm tree. Remove only those fronds below an imaginary horizontal line across the top of the palm tree trunk. If too many fronds are removed, this can lead to disease, injury to the bud where palm fronds grow from, and weakness in the palm tree's trunk that may snap off in high winds. For tall palm trees, hire a certified arborist to prune your palm trees using the proper techniques.

PRUNING SHRUBS AND GROUNDCOVERS

Arguably, the most objectionable maintenance practice is taking a lovely shrub and stripping away its beauty. Let's be clear—shrubs don't grow into balls, squares, or rectangles naturally. Pruning them into these shapes takes away much of their beauty and puts unneeded stress on the plants. While some shrubs such as Japanese boxwood (*Buxus microphylla* var. *japonica*) and dwarf myrtle (*Myrtus communis* 'Compacta') handle formal pruning fairly well, shrubs native or adapted to arid regions don't respond well to repeated shearing.

Shrubs and groundcovers in the garden setting do need to be pruned from time to time to promote their health and beauty, although new plants won't need any pruning for the first year. Many shrubs are pruned once to twice a year in a landscape setting, with some shrubs and groundcover only requiring pruning every three years or more.

Spring Pruning

If you find yourself nervous about pruning your shrubs, I encourage you to relax. Unlike trees, shrubs are pretty forgiving of inadvertent pruning mistakes, and they will serve as helpful lessons to help you become a better gardener.

Dwarf oleander 'Petite Pink' (*Nerium oleander* 'Petite Pink') shrubs that have undergone rejuvenation pruning. New growth will soon follow.

THERE ARE DIFFERENT WAYS TO APPROACH SPRING PRUNING:

- Remove frost-damaged growth from plants in spring once the threat of freezing temperatures has passed. Focus on pruning brown, crispy growth using hand pruners until you reach green, living tissue. Loppers can be utilized for larger branches if needed.

- Before you prune healthy shrubs and groundcovers, determine whether they need pruning. Look to see if they have healthy growth, an attractive natural shape, and room to grow. For plants that match these criteria, step back and determine whether they need to be pruned at all.

- For shrubs that need to be pruned due to limited room, begin by pruning them back to half their size in spring. Use hand pruners, loppers, or even hedge trimmers to do this pruning task. This is an easy pruning method for most flowering shrubs, giving them plenty of room to grow in their natural shape the rest of the year.

- Selective thinning is a pruning practice done in spring, where selected branches are removed from the interior of shrubs to help open them up or remove old, woody branches. This is a popular technique for shrubs when you don't want to significantly reduce the height of shrubs or desire a naturalistic landscape style. Remove one-third of the branches as close to the base of the shrub as you can get using loppers or a pruning saw. Then, the following year, prune another third, and so on, which will remove old, unproductive growth without the need for major pruning.

- Rejuvenation pruning is another pruning tactic that is helpful for shrubs and groundcovers with old, woody growth that produce less foliage and flowers. Many of them can be pruned back severely to 6–12 inches (15–30 cm) above the ground, which stimulates new branches with more leaves and flowers in healthy plants. Loppers and a pruning saw are the best tools to use for this pruning technique for shrubs. Hand pruners or a hedge trimmer are most useful for groundcovers. This form of pruning is done in spring, and it can take up to eight weeks to see new growth. Some shrubs and groundcovers may not grow back from rejuvenation pruning, indicating old age or poor health. If they don't grow back, replace them with new ones. Rejuvenation pruning should never be done to trees!

- In late summer, you may want to lightly prune summer-blooming shrubs and groundcovers if they are getting too large or in danger of outgrowing their space. You can lightly prune by removing one-third of its outer growth or pruning some interior branches. Avoid major pruning this time of year because any new growth will be more susceptible to damage from the winter cold.

REHABBING FORMALLY SHEARED SHRUBS

We know how harmful and ugly the effects of repeatedly shearing your shrubs are. But what do you do if you have shrubs like this in your landscape? The good news is that you can help get them back to their natural form in most cases. Rejuvenation pruning is the method to help revert your shrubs to what they should be. Again, this should be done in spring when temperatures are beginning to warm and plants are ready to put on a flush of new growth. If your sheared shrubs don't grow back from rejuvenation pruning, rest assured that you gave them a chance but that the years of maltreatment were just too much for them. In that case, replace them with new ones and maintain them the right way.

PRUNING ORNAMENTAL GRASSES, PERENNIALS, VINES, CACTI, AND SUCCULENTS

Most pruning in your garden will be centered around trees, shrubs, and groundcovers.

However, we need to mention other plants' pruning requirements, which are slightly different, especially when it comes to cacti and other succulents.

Ornamental Grasses

For people who want beauty with little fuss, ornamental grasses add beauty and texture with low maintenance requirements. Most ornamental grasses need pruning once a year in early spring once freezing temperatures have ended. At that time, prune them back to 3 inches (8 cm) from the ground. Yes, this is severe pruning and should be done annually for the health and attractiveness of the grasses. Hand pruners will suffice for small clumps of grass, but you can use hedge trimmers or a pair of pruning shears for larger grass clumps. Grasses need no other pruning the rest of the year.

Flowering deer grass (*Muhlenbergia rigens*) adds vertical height and texture at the Scottsdale Xeriscape Garden in Arizona.

Perennials

Perennials are generally smaller flowering plants in the arid garden, and their pruning needs are simple. Hand pruners will be the best tool for your perennial pruning tasks. Prune to reduce their size in spring if they become too large. Also, prune to remove flowers as they fade in perennials such as penstemon (*Penstemon spp.*) and angelita daisy (*Tetraneuris acaulis*) or remove unproductive growth to improve their appearance. Perennials can become old and woody and benefit from severe pruning in spring every few years. Perennials don't live as long as other plants, so expect to replace them every few years.

Vines

The beauty vines add to vertical surfaces within our landscapes is undeniable, and their growth habit can make pruning a bit challenging unless you know the right way to approach it. Major pruning of vines should happen in spring, with summer-flowering vines being pruned in early spring once frost has passed. Prune vines that bloom only in mid to late spring once their flowers fade. In summer, lightly prune if needed to prevent vines from outgrowing their space or growing into your neighbor's yard.

Our first goal is to keep vines from outgrowing their allotted area. Vines can grow very tall and wide but can be maintained much smaller. Remove any stems that are dead, unproductive, or growing in the wrong direction. Second, look for highly tangled stems and prune those out as well. Avoid the urge to pull out a mass of stems haphazardly; instead, focus on individual stems for removal. Hand pruners work well to keep wayward tendrils from straying out too far, or you can attempt to weave them back

into the central part of the vine, where they will continue to grow. For vigorous vines that are overgrown, you may opt to severely prune them back to the ground in spring.

Pruning Cacti and Other Succulents

For many types of cacti, periodic pruning is sometimes needed every few years. All pruning of cacti and other kinds of succulents should occur in spring. Branching types of cacti that form multiple stems or pads will likely need pruning every few years. Columnar cacti that grow upright may not need pruning unless it has multiple stems and grows wide.

One of the most likely cacti that will need pruning is prickly pear cacti (*Opuntia spp.*), with their thin, rounded, segmented stems. Over time, new growth typically occurs outward, with new pads increasing their width. In addition, the older pads in the middle can begin to brown and die. Pads that rest on the ground increase the likelihood of rodents taking residence under the cactus. Pruning should focus on removing pads resting on the ground and removing any dead prickly pear pads. You can utilize loppers, or a pruning saw, to make pruning cuts where the pads connect. Use great care to avoid the spines and tiny hairlike glochids present on most prickly pear species, which are painful and irritating. A pair of tongs or a shovel can help you pick up fallen pads. Wear long sleeves and gloves, but don't touch the pads with your gloves as the glochids will ruin them.

OUCH! GETTING RID OF CACTUS SPINES

Dry climates populated with cacti are prevalent, and so is the risk of getting pricked by their sharp spines—I speak from personal experience! There are several ways to get them out of the skin, including duct tape. However, there are two methods that, when combined, work best. First, use tweezers with pointed tips to remove as many spines as possible. Follow that up by applying a layer of white glue over the affected area and placing a layer of gauze. Once the glue dries, pull off the gauze, which should help remove any smaller spines.

Upright cacti may only produce a single stem, like the iconic saguaro cactus, and don't need pruning. However, certain columnar cacti also produce multiple stems like *Cereus* or *Pachycereus*. If they have enough room to grow, pruning isn't required. But, if there are wayward or diseased stems, removing a single stem may be needed. Using a pruning saw, make cuts at the base of the cactus. You can often use a healthy section of the cut stem to plant elsewhere in your garden—see Propagating Cactus section from Chapter 2.

Succulents are a favorite of dry climate gardeners for their unique shapes and ease of care. While most are relatively carefree, some do need pruning occasionally. First, use hand pruners or loppers to remove dead, diseased, or unattractive growth. Many succulents such as

An artichoke agave (*Agave parryi v. truncata*) surrounded by offsets (also referred to as "pups").

The lower pads of this flowering Santa Rita prickly pear (*Opuntia santa-rita*) touch the ground and should be pruned off to help prevent unwelcome rodents from taking up residence underneath.

aloe and agave have a spreading growth habit, producing offsets (or baby) plants next to the parent plant. For agave offsets, you can keep some growing around the parent plant, but avoid allowing all of them to grow to maturity as you will have to deal with an overgrown plant and a haven for garden pests such as rodents and scorpions. Remove agave offsets from the parent plant by cutting the fleshy stem underneath, connecting them. You can replant the little agave elsewhere, gift it to someone else, or discard it.

Like agave, many aloe plants produce offsets, including medicinal aloe (*Aloe vera*) and 'Blue Elf' aloe (*Aloe* × 'Blue Elf'). The offsets (baby aloes) require removal every few years to keep them within bounds and keep the aloe attractive. Untended aloe clumps allowed to grow unchecked become filled with old, woody plants and look ugly. Aloe maintenance focuses on removing any old aloe and excess aloe offsets. A simple way to maintain aloe is to dig up the entire clump every three years, remove a healthy section with roots attached, and replant it in the same spot. Give away, replant elsewhere, or discard the rest.

In spring, prune other succulents to promote their health and appearance. Keep in mind that some succulents have a sap that is irritating to the skin and the eyes or is toxic if ingested. Succulents within the Euphorbia family are primarily known for their milky white sap, which can cause adverse skin and eye reactions and are poisonous if ingested. Wear long sleeves, gloves, and eye protection when pruning all succulents—it's better to err on the side of caution when you aren't sure about the succulents in your garden.

FERTILIZER —WHAT PLANTS NEED IT AND WHY MANY DON'T

In my years of experience helping people with their dry climate gardens, a common mistake is fertilizing plants that don't need it. One characteristic trait of plants native to arid regions is that they don't require highly fertile soil. This adaptation confuses many people who believe that all plants need supplemental fertilizer. However, plants adapted to living in dry climates come from regions with low nutrient levels in the soil and have evolved to survive and thrive without supplemental fertilizer. Fertilizing plants like these often result in a floppy flush of foliage and can even reduce flowering on blooming plants. Fertilizer also makes plants grow faster, increasing the pruning and water required. Feeding plants unneeded fertilizer is another example of complicating things in the garden and creating more work for ourselves. While most arid-adapted plants don't need supplemental fertilizer, some do. But, let's first talk about what fertilizer is.

TYPES OF FERTILIZER

Fertilizer adds supplemental nutrients for plants when the amounts present in the soil aren't sufficient. Nutrients are represented by three letters, NPK, on a package of fertilizer, which stand for Nitrogen, Phosphorus, and Potassium. These are the three primary nutrients that plants need to grow. Micronutrients are nutrients that plants need but in smaller amounts. The type and quantity of major nutrients can vary between fertilizers. There are three numbers on every bag of fertilizer separated by dashes, representing the percentage of each nutrient present in the fertilizer.

Before applying granular fertilizer, clear away any mulch, sprinkle directly on the soil, and lightly rake it into the top couple inches of soil. Then reapply the mulch over the top.

There are two main types of fertilizers: synthetic (manufactured) and organic (natural materials). Synthetic fertilizers are manufactured from inorganic sources and are fast-acting. However, the nutrients are depleted quickly and may need to be applied frequently. Because they leach out of the soil rapidly, they have no lasting beneficial effect on plants or the soil. While easy to apply, it is easy to "burn" your plants by adding too much synthetic fertilizer, which can dry out your plant or even kill it.

Organic fertilizers come from natural materials. They are slower acting and have lower levels of nutrients as opposed to synthetic fertilizers. They release their nutrients over several months, so their effects last a long time and don't have to be applied as often. The chance of them burning your plants is much less than synthetic fertilizers, and over time, they help improve your soil by encouraging beneficial microorganisms that benefit plants.

Natural materials used in organic fertilizer include both animal and plant sources. Animal sources include blood and bonemeal as well as manure. Cow, chicken, and rabbit are popular sources of manure, which are aged over several months before it is suitable for the garden. In addition, fish emulsion and plant byproducts such as kelp from seaweed are well-known organic fertilizers. The labeling of organic fertilizers will tell you the percentage of each major nutrient within the fertilizer and the natural materials used.

Slow-release fertilizers are a convenient and easy way to fertilize potted plants.

Slow-release fertilizers are an excellent tool for the gardener. They release nutrients over several months, reducing the frequency of fertilizing and ensuring a steady release of nutrients for plants. These fertilizers are helpful for container planting but are suitable for most planting situations. Because they release nutrients at a low rate over three to four months, there is little danger of fertilizer burn. You can find both synthetic and organic forms of slow-release fertilizer.

WHY MANY NATIVE DRY CLIMATE PLANTS DON'T NEED SUPPLEMENTAL FERTILIZER

In most cases, plants from arid regions don't need you to fertilize them. It's true! Many dry climate shrubs such as *Caesalpinia*, *Calliandra*, and *Senna*, fertilize themselves by making their own nitrogen. Trees like mesquite (*Prosopis spp.*) and palo verde (*Parkinsonia spp.*) do as well. However, even if your favorite arid native doesn't make nitrogen, it is likely adapted to low nutrient levels in the soil and doesn't need you to add any more nutrients.

PLANTS THAT NEED FERTILIZER

Whether or not you have fertilizer stored in your garden shed will depend on what type of plants you have and where they are growing. If you have plants native to your region or a similar region, you are likely not to need fertilizer. However, many gardeners grow more than arid-adapted plants—some that need fertile soil and supplemental fertilizer in our low nutrient soils.

Tropical hibiscus (*Hibiscus rosa-sinensis*) plants need supplemental fertilizer to grow their best in dry climate soils. The silvery cassia shrub (*Senna phyllodenia*) in the background is native to dry climate regions of Australia and requires no fertilizer.

In dry climate gardens that experience mild winters with a few light freezes (or none), tropical plants like hibiscus are a popular element. But because tropical regions have fertile soil, plants native to those areas will need more nutrients that are present in our soils. So, regular fertilizer applications are likely to be required throughout the active growing season, which is spring, summer, into early fall. The same is true for plants native to a more temperate climate where soils are richer in nutrients. A general rule is that if you have a plant native to a region that isn't an arid or semi-arid climate, chances are it will require supplemental fertilizer. Look for an all-purpose fertilizer that contains equal amounts of nitrogen, phosphorus, and potassium. A good rule of thumb is to apply fertilizer in spring and again in late summer, following package instructions.

Palm trees are a characteristic tree of arid gardens. California fan palms (*Washingtonia filifera*) and Mexican fan palms (*Washingtonia robusta*) are native to desert regions and usually don't need supplemental fertilizer. However, other palm species like Mediterranean fan palm (*Chamaerops humilis*), pygmy palm (*Phoenix roebelenii*), and queen palms (*Syagrus romanzoffiana*) perform best with a palm fertilizer applied in late spring and early summer. Avoid using a general all-purpose fertilizer as palm trees have specific nutrient needs that traditional plant fertilizers don't meet. Palm fertilizers contain high levels of nitrogen, potassium, and certain micronutrients that they need. The fertilizer's effects won't show up in the existing fronds (leaves) when fertilizing palm trees. Instead, the results will show in the future growth of the palm tree. Follow the directions on the fertilizer package and water well afterward to help the fertilizer reach the palm tree's roots.

Flowering annuals need regular applications of fertilizer to help maximize their blooms. A combination of slow-release fertilizer applied every three months and a monthly application of liquid fertilizer with higher phosphorus levels to promote flowering works well. Read the fertilizer label to ensure that they will help promote blooming.

Finally, plants grown in containers or raised beds need fertilizer—even arid natives, cacti, and succulents. This is because plants in containers have no access to nutrients in the soil, so we must regularly provide them with nutrients in the form of fertilizer. Slow-release fertilizers are ideal for container plants as you only need to apply them every few months. Other plants (not discussed in this book) that need regular fertilization include fruit trees, lawns, roses, and vegetables.

APPLYING FERTILIZER

Plants need nutrients most when they are actively growing. For most plants, this is in spring through late summer. Our fertilization schedule revolves around the growing season, which is between your last frost date in late winter (or early spring) and the first frost date in late fall or early winter. The frost date is when temperatures dip below 32°F (0°C), considered freezing. It's safe to fertilize once

FERTILIZER: USING LESS IS BEST

Whichever fertilizer you select, carefully follow the directions regarding how much you apply, as different fertilizers have varying levels of nutrients. Adding too much is bad for plants and can cause "fertilizer burn." When in doubt, add slightly less than the label recommends—especially when using synthetic fertilizers. The label should also instruct you on the best way to apply the fertilizer.

freezing temperatures have ended in spring, and end fertilizing at least eight weeks before your average first frost in late summer or early fall. Plants should need no fertilizing in winter except for plants actively growing like flowering annuals.

When applying granular or organic fertilizers, lightly rake them into the top inch or two (3–5 cm) of soil to allow them to make contact with the soil. Water well before and after fertilizing to help the fertilizer break down and reach the roots. Liquid fertilizers can be applied using a watering can or a hose-end sprayer container. Whichever type of fertilizer you select, apply it over the plant's root zone, where the roots are most present. Roots generally radiate from a plant extending outward to where the branches reach.

A pink and white bougainvillea in a light-blue pot at the Desert Botanical Garden in Phoenix, Arizona. When grown in the ground, it does fine without supplemental fertilizer; however, when grown in a pot, bougainvillea (and any other plant) does require fertilizer as it has no access to nutrients in the soil.

Natal plum (*Carissa macrocarpa*) shrubs with frost damage.

FROST PROTECTION

Many arid and semi-arid regions experience temperatures below freezing in winter, significantly affecting certain plants. Every plant has a minimum cold temperature it can survive, and it varies depending on the type of plant. Some plants are frost-tender and suffer damage when temperatures go below 32°F (0°C). Cold damage occurs when the water inside a plant freezes, expands, and breaks the cell walls resulting in the death of part of the plant. This damage is referred to as frost damage and appears as brown, crispy leaves, usually along the outer part of the plant. Frost damage is cumulative, so the more often the temperature falls below freezing, the more damage the plant will incur. Severe frost damage can cause the death of the plant. Frost-tender plants come from frost-free regions with mild winters and don't experience freezing temperatures. These warm-loving plants

are popular additions in dry climate gardens where freezing temperatures may occur and suffer damage if not protected.

So, what can you do to avoid frost damage? First, if you are using plants native to your region, you don't have to worry about it, as they are perfectly adapted to your seasonal temperatures. Secondly, you can choose plants from regions with similar minimum temperatures to those your area experiences in winter. Finally, you can grow plants native to areas with colder winter temperatures. Young plants almost always benefit from frost protection through their first winter, when they are most susceptible to cold damage.

For cold-winter climates where weeks of below-freezing temperatures are the norm, you can bring potted frost-tender plants

Frost cloth covers frost-tender trailing lantana (*Lantana montevidensis*) in the author's garden. The flowering angelita daisy (*Tetraneuris acaulis*) in the foreground is cold-hardy to -10°F (-23°C) and doesn't require protection from the cold.

indoors until spring. In regions with the *occasional* freezing temperature, you need to make a decision regarding your frost-tender plants. The first option is to do nothing and prune away the frost-damaged growth in spring once the threat of freezing temperatures has passed. This option works if plants are root-hardy, so even if the top part of the plant suffers extensive frost damage, it will grow back from the roots once the weather warms up. However, frost-tender cacti and other succulents need protection from freezing temperatures as they can suffer significant damage and are slow to recover.

Another option for the occasional freeze is to protect your plants with frost cloth, blankets, sheets, or towels when temperatures fall below 32°F (0°C). Be sure to cover the entire plant and leave no gaps between the covering and the ground, or warm air will escape. Covering plants allows your plants to weather freezing temps into the upper 20s (-7°C) relatively unscathed and helps prevent the brown, crispy foliage that follows after a freeze. However, if you have several consecutive days of temperatures below freezing, significant frost damage may occur even if your plants are covered. If using towels or sheets, uncover plants in the daytime when temperatures heat above freezing to allow plants to warm up and get sunlight. Frost cloth allows sunlight to penetrate, so you don't have to take it off in the daytime. For frost-tender cacti, you can use Styrofoam cups to cover the tips of individual cacti stems to protect them from damage from the winter cold.

WATERING

Water is a precious resource in arid or semi-arid regions. We celebrate when it rains and, conversely, are becoming more familiar with ever-increasing periods of drought. Yes, we can create vibrant gardens filled with beauty that will thrive, but we must consider plants' watering needs and find ways to eliminate wasting water.

MOST DRY CLIMATE GARDENS ARE OVERWATERED

Up to 70 percent of a household's water usage goes toward the landscape in dry climate regions. While this amount may seem relatively high, in large part, it's due to people using high-water plants and overwatering their plants. Unfortunately, the average homeowner has no idea how much water their plants need. As a landscape consultant, I have seen this same mistake repeatedly, resulting from misleading information and the general perception that our plants need lots of water since we live in a hot, dry climate. Sadly, this is often not true, and overwatering leads to health problems for your plants, not to mention an expensive water bill.

Why Overwatering Is Bad for Plants

When we overwater plants, it decreases the amount of oxygen present in the soil that they need to survive. Excess watering flushes out nutrients that plants need to live. Iron chlorosis is caused by overwatering when the iron in the soil leaches out from too much water. The younger leaves will turn light green while the veins remain a darker green. Plants with open, lanky growth and a reduction in flowering may signal they are receiving too much water. Other signs of an overwatered plant are leaves wilting or turning yellow. In the case of too much water, the soil will be consistently moist or wet. Finally, plants that get more water than they need will have excessive green growth, fewer blooms, grow overly large, and require more maintenance. Conversely, plants not getting enough water will show dull green or browning leaves, new growth that wilts during the day, and the soil will be pretty dry.

Why Deep and Infrequent Irrigation Is Best

When we water plants, the goal is to give them the water they require to remain healthy while helping them to grow deep roots. We want the roots to grow deep because the soil is cooler

Extreme example of overwatering where water runs off the sidewalk and into the street. You can be watering too much even if you don't see any runoff.

further down, and it holds onto moisture longer, which means less frequent watering is needed. If you water plants too shallowly, the roots remain near the surface, where they are susceptible to hot summer temperatures, and the soil dries out much faster. In addition, arid soils have naturally occurring salts, and shallow watering allows the salts to concentrate around the root zone. Too much salt isn't good for plants, and signs of salt damage may appear on the leaves in the form of brown leaf tips or edges. Another sign of excess salt buildup is a powdery white substance around the soil surface. Deep watering helps flush out naturally occurring salts and keep them away from the roots of your plants.

RECOMMENDED WATERING DEPTHS IN INCHES	
Trees	20–36" (51–91 cm)
Shrubs/Vines	18–24" (46–61 cm)
Groundcovers	12" (30 cm)
Cacti/Succulents	8–12" (20–30 cm)
Lawn	6–10" (15 cm)

The table above is a helpful reference guide to help you determine how deeply water should permeate the soil. As you can see, the larger a plant is, the deeper it needs to be watered. To determine how deep you are watering your plants, use a soil probe and push it into the soil after watering—alternatively, you can use a long piece of rebar in place of a soil probe. It should go down into the ground and stop when it hits dry soil, giving you the depth water has reached. Adjust the length of time you water until you attain the appropriate depth. Once you determine how long you need to water, the length of time you water won't change during the year, only the frequency.

TYPES OF WATERING SYSTEMS
There are different types of watering systems popularly used in dry climate gardens. They vary in how they deliver the water, how efficient they are, and their ease of use. I recommend having an irrigation system that connects to a timer or irrigation controller, making caring for your garden more manageable and ensuring reliable water delivery to your plants. Water is the most crucial factor in the garden as

to whether plants will survive or not, and an automatic irrigation system ensures that there will be no disruption. However, if you only have native plants in your landscape that need little to no supplemental water, you may not need an irrigation system installed.

Let's discuss commonly used irrigation systems within dry climate gardens to help you choose the best one for you. You can have it professionally installed or do it yourself with whichever irrigation method you decide to use. Many helpful resources, including classes offered by local water utilities or your botanical garden, can be invaluable resources to help you learn how to set up your irrigation system.

Drip Irrigation
While this is the newest irrigation method, it is the most efficient form of irrigation for the dry climate garden. Water flows through underground tubes that lead to small ¼-inch (6 mm) tubing and then through emitters that release droplets of water to the base of plants at a measured rate. Standard drip emitters vary in

WEATHER-BASED IRRIGATION

Irrigation controllers have come a long way in helping you maximize watering efficiency. Many models monitor current weather conditions and automatically adjust watering schedules. Referred to as "smart controllers," they take the guesswork out of how much water plants need and when to water.

A drip irrigation emitter slowly drips out water, watering this shrub to 18 inches (46 cm) without danger of runoff.

the amount of water they release per hour. Popular sizes range from 0.5 gph (gallons per hour) (2 liters per hour) to 4 gph (15 lph). In addition, you can adjust the amount of water that each plant receives by using adjustable drip emitters. Drip irrigation is the most efficient irrigation method as water is slowly delivered straight to the roots zone of plants with little to no water lost to evaporation. It also allows water to soak into the soil without fear of runoff.

Place drip emitters around the dripline of plants, where the roots of a plant are concentrated. A plant's dripline is where the leaves extend out to, or in other words, at the outer reaches of the plant's branches. As a plant grows wider, you should move emitters outward to keep up with the expanding dripline and the root growth. Remember, roots don't remain in a small area; they grow outward as the plant does. As plants grow larger, they need more water, so you will need to add more emitters as they mature.

Drip emitters release varying amounts of water depending on their size. The bigger the plant, the more water it will need. You can purchase emitters that emit specific amounts of water over an hour or select an adjustable emitter that you can adjust the amount of water that comes out. Here are recommendations for drip emitter size for different types of plants: **Trees**—4 gph (15 Lph) **Shrubs/Groundcovers/Vines**—2 gph (8 Lph) **Perennials/Small Plants**—1 gph (4 Lph) **Cacti/Succulents**—0.5 gph (2 Lph).

Recommended placement for drip emitters around a tree's drip line.

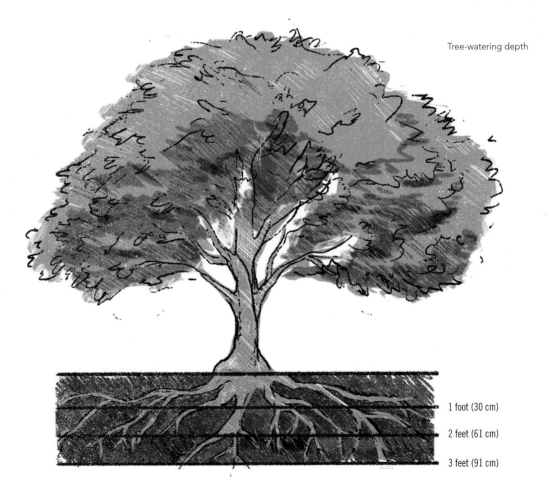

Tree-watering depth

1 foot (30 cm)

2 feet (61 cm)

3 feet (91 cm)

Bubblers

Bubblers are commonly used to water trees and release a ½ gallon to 2 gallons (2–8 L) of water every minute. They help cover large root zones of plants such as trees or large shrubs. However, they aren't the best option for plants with differing water requirements in the same area, as they do not adjust to individual needs. Water is released quickly through a bubbler that rests above the ground and wets the entire root zone. Use a soil probe to determine how long to water to reach the optimal depth. Because they release water quickly, it is best to use them in areas with a border or basins to avoid the water running off. Bubblers have a shorter run time than drip systems because they release water quickly.

Sprinklers

Sprinklers are a common sight for landscapes with a lawn, but they are also sometimes used to water plants. If you have plants within a foot or two (30–61 cm) of your lawn's sprinkler system, you may not need any separate irrigation for the plant as it will get what it needs from its proximity to the lawn. However, sprinklers aren't the most efficient way to water plants as it is harder to target where the water goes to individual plants. In addition, when sprinklers spray water, up to 50 percent is lost to evaporation right away—especially on a hot, dry summer's day. If you currently use sprinklers to water your plants, you can transition over to a drip irrigation system for your plants, which will help you to meet their water needs and help you use them as efficiently as possible. If you use sprinklers within your landscape, keep desert native plants and trees out of reach of any overspray as they suffer problems when kept too wet.

DO YOU USE YOUR LAWN?

Grass lawns are thirsty and use three times more water annually than a swimming pool and up to ten times more water than a landscape filled with native and desert-adapted plants. If you have a lawn, make sure it provides a function such as a play surface for your children. However, if you like the look of a lawn but rarely use it, look to replace it with drought-tolerant groundcovers with lush green foliage that require a fraction of the water.

'Outback Sunrise' emu (*Eremophila glabra* 'Mingenew Gold') planted in a mass to create an area of lush green in the landscape.

Watering by Hand

Many people like to water their plants by hand, using a hose or watering can. However, watering your plants with a hose isn't always as effective as other irrigation methods. One of the problems with using a hose to water your plants is that the water comes out so quickly that it begins to run off before it can permeate the soil. This is especially true for clay soils, which take longer to absorb water than sandy soil does. The result is a shallow-rooted plant susceptible to salt buildup and is more apt to experience stress when hot, dry weather arrives.

Another issue with hand watering is that it can lead to an uneven watering schedule. It is easy to skip or forget to water your plants if you get busy. One of the benefits of having an irrigation system connected to an irrigation clock (or controller) is that you can set it and forget it. In addition, you will have the freedom to leave on vacation with no worries about your plants getting watered.

While relying on hand watering as your sole irrigation method isn't the easiest or most efficient way to water, it is handy in helping to rectify problems with salt buildup in the soil. Put your hose along the plant's dripline and turn it on a very slow trickle for at least an hour or two. For large shrubs or trees, you will need to move the hose around the dripline until the root zone around the dripline has been deep watered. This extra influx of water will help to flush the salts deep beyond the root zone. Alternatively, you can attach a soaker hose to the hose and wrap it around the plants needing extra water. Doing this once is usually enough to flush out the salts as long as you set your irrigation system to water deeply moving forward.

Citrus tree with a bubbler irrigation head. The powdery white substance around the root zone is a buildup of salts due to shallow watering.

Rainwater Harvesting

The purest water comes from rain, and it's free! Rain can be funneled from your roof into a rain barrel or cistern, where it is ready for you to use when you need it. Gutters around your roof can connect to a rain barrel. To utilize rainwater, connect your hose to the hose bib at the bottom of the barrel and water your plants. In addition to storing rainwater in a rain barrel, you can contour your landscape to direct water to specific plants—see Rainwater Harvesting section in Chapter 3. While rain in arid regions can be sporadic when it falls, why not use it for your plants instead of allowing it to run off into the street?

THE BEST TIME OF DAY TO WATER PLANTS
Water plants in the early morning hours before the sun rises, allowing plants to take up water before the day's heat arrives. Plants quickly absorb this early morning drink, so they are ready to face the day. Avoid watering in the middle of a hot summer day, as plants have difficulty taking up the water and won't use it efficiently. A common mistake is using sprinklers to water a lawn or plants in the middle of the day, resulting in much water being lost to evaporation before it hits the ground. We need to make the most of our water and reduce waste, so early morning watering is best. In general, watering at night or in the evening isn't considered ideal as it can foster the formation of fungal diseases in plants.

HOW OFTEN TO WATER

The most common question I get asked is, "How often should I be watering my plants?" The answer is that it depends. The frequency of watering plants is dependent on the type of plant, whether the soil is clay or sandy, if it's a new or mature plant, and the time of year. In addition, dry climate plants don't like constantly moist soil, so a good guideline is to allow the soil to dry out somewhat between watering.

WATERING REQUIREMENTS BY PLANT TYPE
As discussed earlier in this chapter, not all plants need the same amount of water or to be watered at the same frequency. The smaller the plant, the more often it needs to be irrigated because it has a smaller root system. Trees require more water, but with their extensive roots, they don't need to be watered as often as plants with smaller roots systems. Cacti and other succulents don't need frequent watering because they store water inside their leaves and stems. Native or arid-adapted plants require less regular watering than those from different climates.

Below is a list of plants in order of watering frequency, beginning with those that need to be watered more often to less often.

Flowering Annuals
Lawn
Vines/Groundcovers
Shrubs
Trees
Cacti/Succulents

Unfortunately, many watering systems consist of a single irrigation line for all types of plants within the landscape. This means that the trees, shrubs, and groundcovers are all on the same watering schedule, and you need to set the schedule for the plants that need water most often. As a result, frequent watering can lead to problems with trees, which grow too fast and form weak wood, leading to branch breakage, less oxygen in the soil, and an increased incidence of fungal disease. Cacti and other succulents also suffer from too much water. Thankfully, we can correct the problems caused by all plants being on the same watering schedule.

Irrigating by Zones

Configure irrigation systems so that plants with similar watering needs share their own zone. Irrigation zones are separate areas connected to a particular irrigation valve programmed with a specific run time and schedule. Irrigation controllers can operate different zones, so the individual water needs of each type of plant

New drip irrigation system installation in a front yard that will include multiple lines to meet the needs of various types of plants.

are met. For example, a garden filled with trees, shrubs, and groundcovers would need two separate zones—one for the trees and the other for shrubs and groundcovers since they share similar water requirements. If you have a mixture of native/arid-adapted plants mixed with higher-water use plants, separate them with different irrigation lines to maximize your water use efficiency. I like to add an irrigation zone for potted plants when possible. Finally, if you have a lot of cacti and other succulents that need supplemental water, you can add a zone for them

too! The more we correctly meet the watering needs of our plants, the healthier they will be.

If you need to manage by using a single zone for all your plants, there are some strategies you can employ to meet the individual needs of your different plants if you use drip irrigation. Use adjustable emitters to modify the amount of water each plant receives. The emitters can be turned off in response to the recommended schedule for each type of plant.

OTHER VARIABLES THAT AFFECT WATERING FREQUENCY

In addition to the type of plant, we need to consider other variables to determine how often to water our plants. For example, a significant factor that affects watering is soil type. Clay soils are slow to absorb water, and they dry out slowly. In contrast, water penetrates sandy soils quickly and dries out fast. Therefore, when it comes to irrigation frequency, plants in clay soil need to be watered less often than those in sandy soil. Older irrigation systems can become less efficient over time and develop leaks, so they should be evaluated every few years to see if replacement is needed.

New plants require more frequent watering than those that are established. Extra water is needed because young plants have a much smaller root system than mature plants. During their first year, plants grow new roots, but they will need water more often throughout the process until they are established. Trees take longer to establish, typically two to three years. If you have a lot of new plants, you can change your irrigation schedule to fit their needs. However, if you have a few new plants planted among mature ones, you can give the new ones an extra drink of water with your hose in between watering cycles.

Finally, let's talk about the seasons and how they affect your irrigation schedule. Your irrigation system shouldn't be on the same schedule all year. For example, a plant doesn't need as much water in winter as in summer. But, if you continue to water plants in winter as often as you did in summer, you will run into problems as plants aren't growing actively in winter, and their water needs are much less.

PORTABLE DRIP IRRIGATION

Create your own DIY drip irrigation system for plants that need infrequent irrigation. Take a clean 1-gallon (4 L) plastic jug and poke five holes in the bottom with a nail. Fill with water from the hose and place next to the plant you want to be watered. The water will slowly drip out, giving your plant the water it needs.

In addition, doing so invites the risk of rot within the plant or the roots.

How often you water needs to change with the seasons. In summer, you will water most often, and in the winter, much less—watering frequency in spring and fall midway between the frequency of summer and winter. So, you should change your watering frequency each season. Get to know how to operate your irrigation controller (timer), so you can make the changes yourself, or you can ask a landscape professional to do it for you. If you have a "smart irrigation controller," they can make the schedule changes automatically.

Watering in a Heat Wave

While heat waves are nothing new in the dry climate garden, they are increasing in intensity and frequency. Plants can and do struggle when temperatures soar above normal ranges. If your garden consists of arid-adapted plants from hot regions, they will weather the average heat reasonably well. Still, they can use a little extra help in a heat wave, as do plants that may be less ideally suited for arid regions. In preparation for an upcoming heat wave, giving your plants a little watering boost the night before is helpful. If you have an irrigation system, this means turning it on manually and watering for an extra hour or so—this is in addition to their regular watering schedule. You don't need to do this every evening—just the night before a heat wave begins. An additional helpful tactic is to lightly mist your plants with the hose each evening, which helps to cool them off. Remember, water is the best tool to help plants through a heat wave.

WATERING SCHEDULING

So, how often should you water your plants? As we discussed earlier, there are many variables to consider. Knowing how often I water the plants in my Phoenix-area garden with its clay soil won't help a person living in Palm Springs, where sandy soil is the norm, or help someone who gardens in an area with less intense heat in summer. Thankfully, the information you are looking for is available from local sources such as your water utility provider or local government for scheduling your irrigation. This region-specific information is particularly helpful because it considers the soil type, temperatures, and weather patterns unique to where you live. *A word of caution—there is much misinformation about irrigation frequency spread by gardeners and even landscape professionals who tend to recommend watering too often and shallowly.*

It's important to remember that guidelines are just recommendations, and you may need to make adjustments within your garden. Look for signs of overwatering (yellowing leaves, iron chlorosis, constantly moist or wet soil) or underwatering (dulling of shiny foliage, wilting leaves, browning foliage, dry soil) and adjust the frequency if needed. If you find that you have been watering too often (which is likely), slowly wean your plants to the recommended irrigation schedule from local resources over several weeks to allow plants to become accustomed to their new schedule. You will likely be watering less often, but the length of time your water is on will be longer to foster deep root growth.

If information isn't readily available, you can determine the best watering schedule for your garden through trial and error. For example, you may begin by watering three times a week in the summer (more often if you have sandy soil). Look for signs of over- or underwatering and adjust the frequency as needed. Perhaps you can stretch out the intervals of watering. Don't be afraid to try different schedules, but be sure to observe your plants to increase or decrease the gaps for the plant's optimal health. Remember, you will need to water less in spring and fall and much less in winter. Be sure to water each plant at the appropriate depth..

Plants native to your region will need less supplemental water than nonnative plants. For example, established desert trees can exist without supplemental irrigation but do appreciate monthly deep watering in summer in times of

Speckled foliage of Baja fairy duster (*Calliandra californica*) due to an infestation of spider mites. Healthy growth is bright green.

animals such as deer, javelina, rabbits, rodents, etc. The good news is that if you use native plants, you will likely suffer less plant damage than you do if you introduce nonnative plants—which garden pests tend to find tastier. Native plants are not pest-free, but there are natural predators, which help to keep their numbers under control. Arid and semi-arid climates benefit from experiencing fewer issues with damaging insects than areas that receive more rainfall. In other words, the more rain, the more pests.

DAMAGING INSECTS

The insects that damage plants in dry climate gardens include agave snout weevil, aphids, caterpillars, cochineal scale, and spider mites. You may find yourself apprehensive when you see insects or the damage they cause to their plants and worry for your plant's health. However, in many cases, your concern is unwarranted. Most plants are programmed to handle a certain amount of damage without seriously affecting their vitality. Often, the insect pest shows up, followed soon after by predatory insects who will come to eat them and limit their numbers. It helps to remember that plants are part of the natural world and need less help than you think to survive damaging insects. However, sometimes plants do need us to step in and help them to control or rid them of damaging insects, as in the case of agave snout weevil. At right is a list of common dry climate insects and how to manage them.

Again, most pest problems are minimal for healthy plants. When damaging insects first appear, the "good" bugs will soon prey upon them and reduce their numbers.

drought. A popular reason for using native plants is not having to provide supplemental water for them. However, that is changing due to increasing temperatures, heat waves, and periods of drought, and you may need to provide additional water from time to time. Keep an eye out for any drought stress in your plants and water if needed.

PESTS IN THE ARID GARDEN

Plants are part of the natural ecosystem and can become affected by unwelcome garden pests. These pests include insects and other

PEST	SIGNS	TREATMENT
Agave Snout Weevil	Wilting agave leaves, sometimes accompanied by a foul odor, resulting in total collapse and death of the agave.	Prevent by drenching soil around agave with insecticide that contains imidacloprid in spring and fall. Remove infected agave and the soil where they were planted. Avoid planting agave in areas previously infected.
Aphids	Soft-bodied insects cluster around new growth, wilting of new leaves.	Spray off with strong jet of water or spray with insecticidal soap.
Caterpillars	Chewed leaves, small holes in foliage, skeletonized leaves, black pellets on flowering shrubs.	Use Bt (Bacillus thuringiensis) to treat severe infestations.
Cochineal Scale	Scale insects suck the juices from prickly pear cacti pads and form a white cottony mass that covers them.	Spray cottony masses off with a strong jet of water, then treat the surface of each pad with insecticidal soap.
Oleander Leaf Scorch	Bacterial disease spread by the glass-winged sharpshooter insect causes outer edges of leaves to turn brown or turn the tips of leaves brown.	To date, there is no known cure. It appears to affect older oleanders and the disease takes two to three years to progress until the shrub dies.
Palo Verde Borers	Large 4- to 6-inch (10–15 cm) black beetles create 1-inch-wide (3 cm) holes as they emerge from underneath palo verde trees in summer. Large branch dieback is another sign. Their larvae feed on the roots.	A healthy tree is the best defense. Water deeply monthly and avoid over-pruning. Insecticides are not very effective.
Snails and Slugs	The difference between snails and slugs is that snails have a shell. Both cause similar problems with chewed leaves as well as slimy trails.	Wrap the base of trees and shrubs with copper strips, which keeps them away. A shallow dish filled with beer causes them to fall in and drown.
Spider Mites	Reddish-brown mites that are related to spiders. These tiny mites suck the sap of foliage, causing speckling of leaves, small white webs, and rust-colored pustules.	Spray foliage in the evening with a hose every few days to dislodge the dust they congregate around, remove badly infected foliage, and spray undersides of leaves with insecticidal soap.
Thrips	Tiny insects that feed on foliage and flower buds. Wrinkled leaves, silvery-green scars on foliage.	Damage is mostly cosmetic and healthy plants usually need no intervention. Insecticidal soap can help early infestations.
Whiteflies	Small, white flies suck the sap from leaves of plants, usually found underneath leaves.	Spray with a hose every few days to remove eggs and young pupae. Use insecticidal soap for severe infestations.

HIDDEN DANGER OF DIY INSECTICIDAL SOAP

Insecticidal soap is one of the most helpful pest control methods for many insect pests and consists of liquid soap and water. The soap breaks down the protective coating of soft-bodied insects and dehydrates them. However, don't make insecticidal soap at home by mixing dish soap and water because it contains strong detergents that can harm plants—play it safe with ready-made insecticidal soap. Also, avoid applying it when temperatures go above 90°F (32°C) as it will burn the foliage of your plants.

DEER, JAVELINA, RABBITS, AND RATS

Animals that are plant eaters see your garden as a feast laid out for them. They aren't trying to ruin your garden—they are just looking to eat to survive, and your plants are more flavorful than what is available in their natural surroundings. However, the damage they leave behind by eating the flowers, fruit, foliage, roots, and cactus pads of your favorite plants is frustrating. Homeowners who live near natural areas within their community are more likely to receive visits from these animals than those in more urban settings.

Add Resistant Plants

It is important to note that it is difficult to prevent these visitors from eating your plants, but there are some things that you can do to help minimize the damage. The first strategy is to look for plants that don't taste good. These plants are "resistant" to being eaten by animals and often include plants with aromatic foliage. *Resistant plant lists are not foolproof.* If an animal is hungry enough, it will eat just about anything, even so-called resistant plants—this can include spiny cactus and succulents. Javelina are especially known to eat just about every plant there is. Native plants are less likely to be eaten than plants brought in from other regions, and they can handle damage better than nonnatives.

Other Pest Control Strategies

The most effective way to keep unwanted animal visitors from eating your plants is by building walls or fencing. A common strategy is to confine tasty plants to a fenced-in backyard or a courtyard where animals have difficulty accessing them. Smaller animals like rabbits and rats can get through fences or sneak under a gate, so wire mesh can help to prevent that. A block wall is very effective in keeping larger animals away.

New plants in the garden are like candy to a hungry plant-eating animal because their leaves are more succulent and tender; therefore, they taste better than those of an established plant. Sometimes, you may need to provide temporary protection for new plants, wrapping wire mesh, such as chicken wire, around the plant and supporting it with stakes. Bury the wire mesh about 12 inches (30 cm) deep to prevent animals from digging under it.

DEER- AND RABBIT-RESISTANT PLANTS

Agave (*Agave spp.*)

Fairy Duster (*Calliandra spp.*)

Black Dalea (*Dalea frutescens*)

Texas Mountain Laurel (*Dermatophyllum secundiflorum*)

Red Eremophila (*Eremophila maculata*)

Turpentine Bush (*Ericameria laricifolia*)

Euphorbia (*Euphorbia spp.*)

Chuparosa (*Justicia californica*)

Lantana (*Lantana spp.*)

Lavender (*Lavandula spp.*)

Texas Sage (*Leucophyllum spp.*)

Oleander (*Nerium oleander*)

Rosemary (*Rosmarinus officinalis*)

Salvia (*Salvia spp.*)

Cassia (*Senna spp.*)

JAVELINA-RESISTANT PLANTS

Turpentine Bush (*Ericameria laricifolia*)

Euphorbia (*Euphorbia spp.*)

Lantana (*Lantana spp.*)

Creosote (*Larrea tridentata*)

Lavender (*Lavandula spp.*)

Heavenly Bamboo (*Nandina domestica*)

Oleander (*Nerium oleander*)

Rosemary (*Rosmarinus officinalis*)

Jojoba (*Simmondsia chinensis*)

Vinca (*Vinca spp.*)

Chicken wire protects a new plant from hungry rabbits in a desert garden.

Other ways to keep unwelcome visitors away include animal-repellent sprays with varying effectiveness. Some contain coyote urine and other odors offensive to animals. In my experience, they are marginally effective and must be applied regularly. Keep your garden tidy; pick up fallen fruit right away and prune overgrown prickly pears cacti that drape toward the ground that animals may find to eat.

Finally, have realistic expectations. If you expect your garden to be a place of perfection, you will likely experience disappointment. Accept that you may have some animal damage from time to time. Do what you can to minimize it and substitute plants that are frequent targets of hungry animals with those that are less appealing.

Alongside a patio, pollinator plants allow close-up views of bees, butterflies, and hummingbirds.

CHAPTER 4

DRY CLIMATE GARDEN DESIGN— WHAT PLANT GOES WHERE?

The design of your outdoor space is a wonderful opportunity to create a garden that uniquely suits you. It isn't that different from what you do inside your home to make it yours. Increasingly, people are looking to the outdoors to expand their living area, and it makes sense as spending time outside decreases stress levels and allows you to fully enjoy all your house and garden have to offer. While many landscape design elements apply to all types of gardens, we need to incorporate the unique features of dry climate design to create a lovely and functional space.

If you have lived in a dry climate for any time, you've likely heard the term "xeriscape." However, there is some confusion about what a xeriscape is, with some people mistakenly referring to it as a "zero-scape," meaning no plants. The origin of xeriscape began in Denver, Colorado, during a drought when people were encouraged to decrease water use around their homes. Xeriscape is created from the Greek word *xeros*, meaning "dry" and it translates to dry landscape. A true xeriscape is a beautiful landscape composed of low-water-use plants, native or adapted to the climatic conditions where they are planted. Organic or inorganic mulches are incorporated around the garden to reduce water loss. As a result, xeriscapes not only use less water but take less maintenance and are resilient to the challenges of dry climate gardening.

The author's front garden in late winter with 'Blue Elf' aloe and angelita daisy (*Tetraneuris acaulis*) in bloom.

Baja fairy duster (*Calliandra californica*) is an excellent choice for full sun, including against south- or west-facing walls where hot, reflected heat is the norm.

PLANT PLACEMENT

In the first three chapters, we have established a great base that you can use to help inform your plant choices and their requirements, which play a large part in the success of your garden. When you ignore these critical factors, the resulting design suffers. It is easy to get caught up in your vision of how you want your landscape to look and place plants solely based on their appearance. Sadly, this is a practice that both homeowners, and some landscape designers, fall victim to as, in many instances, plants struggle as their sun exposure and mature size aren't factored into the design process. As a landscape consultant, I have met with countless homeowners who spent a lot of money on having their landscape professionally designed only to discover that plants ill-suited to their location die or perform poorly. So, it is vital to design with a plant's needs in mind.

It's also important to note that not all plants look or act the same from garden to garden. This is especially true when the same plant is grown in different regions as different factors affect their size, appearance, flowering ability, etc. In addition, variables such as high and low temperatures, wind, humidity, sun intensity, soil, and more all come into play in how a plant responds. So, keep these in mind when comparing what your plants look like to those growing in other conditions.

PLANT EXPOSURE

In the section How the Sun Affects Plants in Chapter 1, we've talked about how a plant's sun exposure plays a large part in the success of how well it will do. So, before designing the area around your home, we need to determine the orientation of the places where you want to add plants—specifically, what direction the area faces. For example, a wall on the east side of your property faces west and receives afternoon sun, and a north wall looks to the south. In addition, trees and other structures can affect where the sun's rays hit, so observe where sunlight falls throughout the day. Once you have this information, match your plant selection according to the exposure where they do best.

ALLOWING ENOUGH ROOM FOR PLANTS

A key component to the success of a well-designed landscape is to ensure that there is enough room for plants to grow to their natural shape: these shapes and other characteristics are how plants add beauty to the area around your home. Of course, the spacing between plants is up to you. A feature of arid gardens is allowing ample space in-between individual plants. Still, you can mass plants together so their outer branches/stems touch (while allowing them to grow to natural shape)—the choice is yours.

GROUP PLANTS ACCORDING TO WATER NEEDS

In dry climates, most plants will need supplemental water, and we need to incorporate this factor into the planning of your garden. Will plants have

SAVING MONEY AND LOOKING BETTER

A good landscape design ensures that plants have enough room to grow to their natural shape and size, which equals an attractive garden. Sadly, the opposite often occurs with over-planting and excessive maintenance that strips plants of beauty. Ironically, planting less will lead to a more attractive garden that takes less time, water, and money to maintain. In this case, less is more.

access to a water source or an irrigation system? If not, consider using native plants or cacti and other succulents that will only need periodic watering that you can do with a hose. If manually watering plants, it is helpful to group plants together with similar watering requirements. You can focus on using higher-water use plants near your house where they are easily reached by a hose and place low-water-use plants further out to reduce the frequency of hauling your hose over a long distance. I highly recommend using drip irrigation systems that allow for separate lines for each type of plant—see the Drip Irrigation section in Chapter 3. A tailored irrigation approach will enable you to intermingle different plant types, meeting their particular watering needs, but don't place cacti or low-water succulents near high-water use plants as they may die from overwatering.

Native and arid-adapted plants create an attractive backyard landscape around the author's home.

LOOK UP BEFORE PLANTING

In regions where rainfall is sporadic, rain gutters may be absent. Instead, "scuppers" are often integrated into strategic places around the eaves of a home that channel water so that it spills downward. Be careful what you plant underneath scuppers or where your rain gutters drain out. Succulents and other plants can struggle due to overly wet soil in these localized areas. Shrubs and perennials may suffer branch breakage due to the force of falling water. A substitute for plants in these areas is to add riprap, large river rock, or decorative containers.

WHAT PLANT WHERE?

Creating an overall plan for your landscape can feel overwhelming, but it doesn't have to be. The key is to take it step by step, and in the end, you'll end up with a beautiful yet functional landscape designed to thrive in an arid climate. We have talked about the first steps of adding plants to your garden—plant exposure, room for plants to grow, and placement according to water needs. The following steps are the fun part of putting together a design—deciding where to put plants. First, we start with the most prominent types of plants, and smaller plant placement flows out from that.

TREE PLACEMENT

Trees are the most significant garden component with their stature and shade. They also provide a visual anchor for the rest of the landscape. To determine the placement of a tree, we must look at other factors, beginning with the distance from the house, other buildings, fencing/walls, patio/decking, sidewalk, swimming pools, and the street. A tree's roots typically extend several feet past its branches, and the roots can cause problems with uplifting and cracking. When selecting a tree, research its mature size and allow enough room for that outward growth so that it won't reach those structures. A common mistake with young trees is failing to realize how big they will become until it is too late. Trees come in all shapes and sizes, so select a smaller type of tree if you want trees closer to those structures.

Utilize trees to create shade, such as over a child's play area or outdoor seating area. If you have plants that do best with filtered shade, place a tree nearby to shade them, especially in the afternoon when the sun's rays are strongest.

The branches of a 'Desert Museum' palo verde tree (*Parkinsonia* x 'Desert Museum') tree provide welcome shade from the afternoon sun for this west-facing house.

Trees can also create shade and cooling in summer to the interior of our homes, mainly when used to block the west- and south-facing sides, which also helps to decrease your energy bill. Keep in mind that many dry climate trees are thorny, so keep those planted away from pathways, the patio, or other areas where people may accidentally get pricked.

From a visual standpoint, trees should be used to frame a structure or shield an unwelcome view. Often, people center a tree in front of a window, which blocks the house. Instead, place the tree just off to the side where its branches will frame the window and your view. A tree should be in scale with the size of your yard. For example, if your backyard is relatively compact, use smaller trees, or a large one will dwarf it. The same is true of the front yard.

WHERE AND WHAT YOU PLANT AFFECTS YOUR NEIGHBORS

When deciding what kind of tree to select and where to plant it within your landscape, consider that these decisions directly impact those whose property borders yours. A tree's canopy can extend over to your neighbor's, adding unwanted plant debris or unwanted shade. Keep in mind the mature size of trees and note how it may affect others. In tight situations, select low-litter trees with a columnar canopy or locate trees with wide branching farther away from property lines.

Out of all plants, trees create the most litter with falling leaves, flowers, and seedpods. First, there are few no-litter plants, and all trees produce some debris, but some produce more than others. Over the years, people have become accustomed to "clean" landscapes, hence the popularity of constant leaf blowing. However, this isn't natural, and we need to get over the idea of a garden almost as clean as the inside of our homes. We can, and should, utilize lower-litter trees near hardscape areas such as your front entry, patio, and swimming pool, where tree litter can accumulate.

SHRUB AND VINE PLACEMENT

Shrubs come in all shapes and sizes and can easily be incorporated into the landscape. Large shrubs are utilized best along bare expanses of wall, along your house, block wall, or fence to break them up visually. Many types can be trained into a small tree form, making them suitable for spots too narrow for regular trees to grow. Tall shrubs can help provide a windbreak along the periphery of your property and delineate property lines. Smaller shrubs make great fillers for bare areas within the landscape, mainly when grouped in threes or fives. They also make an excellent option for use underneath a window without blocking the view and at the base of other structures to provide softness and anchor them visually.

While shrubs can be planted closer to your house than trees, it is still wise to keep them 3 feet (1 m) away to help keep irrigation water from reaching the foundation of your house and provide access to walking behind them if necessary. As with trees, select shrubs that relate to the size and scale of your home and outdoor space—large shrubs for extensive areas and smaller for more restrained landscapes.

Vines can be utilized interchangeably with large shrubs to create vertical interest along fences and walls. One benefit of vines is that they don't take up as much space, making them an excellent choice for narrow spaces and planting beds. There are two types of vines. The first needs a trellis or other support to grow upward. Trellises range from inexpensive wood trellises to decorative, metal ones. Vines aren't particular and will grow up on a handmade trellis. The second kind of vine grasps onto vertical surfaces and grows without support. While this can be handy, the vines may also cause etching or unsightly marks to the surface they grow upon, so be aware of the possibility before using them. Be careful not to plant vines too close to a tree or tall shrub as they can grow into and over them.

Long-blooming shrubs like Blue Bells™ Emu (*Eremophila hygrophana*) add vibrant color and textures to an Arizona desert garden.

GROUNDCOVER PLACEMENT

When putting together a landscape plan, we begin with the most prominent plants before moving toward smaller ones, which are plants like groundcovers that grow low to the ground, averaging 1 to 2 feet (30–61 cm) in height or lower. As their name suggests, groundcovers are utilized to blanket areas of the garden and are usually planted in groups. In dry climates, gardens are covered in landscape rock; groundcovers are instrumental in reducing the heat island effect as they prevent the ground underneath them from heating up. In addition, groundcovers are an excellent visual substitute for a thirsty lawn when planted in mass. Many arid-adapted choices are available, from those with lush green foliage, colorful bloomers, and even low-growing succulents suitable for use as a groundcover. Shorter ornamental grasses make excellent groundcovers, while the taller ones (over 3 feet or 1 m) can be used to substitute for shrubs.

Because of their short height, groundcovers are best placed in front of taller plants like trees and shrubs where they can be seen. You want to use them on the corners next to the driveway, alongside

The lush green foliage and pretty flowers of pink bower vine (*Pandorea jasminoides*) decorate a vertical wall in a courtyard.

Yellow dot (*Sphagneticola trilobata*) is a great visual alternative to a lawn where its lush green foliage and yellow flowers add beauty in both sun and shade.

the entry path, and around trees in the front yard. Groundcovers also make a great pairing next to succulents and boulders. When selecting the right groundcover for your garden, note the mature width, as many can spread several feet.

SUCCULENT PLACEMENT

People who live in an arid region are fortunate to grow succulents in their gardens. It's important to note that cacti are a type of succulent. The unusual shapes of these plants create opportunities for adding unique interest throughout the landscape. Another benefit of using them is that they are genuinely low-maintenance and thrive in dry climates.

There is a misconception that succulents don't need any water, and this isn't true. Native cacti and other succulents may do fine in your garden with regular rainfall. Still, those from different regions that may receive more precipitation will need supplemental water—especially in the summer months.

Succulents can be used in a variety of ways in the landscape. There are countless different shapes and sizes to play with when choosing succulents, from tall, columnar cacti to those that are only a few inches in height. Columnar cacti can be placed in similar locations as trees to serve as a central focal point or interspersed throughout the garden to create multiple vertical visual statements. Medium-sized succulents such as agave (*Agave spp.*) and desert spoon (*Dasylirion wheeleri*) have attractive rosette shapes that make an excellent filler for bare areas. And, let's not forget about prickly pear cacti with their distinctive paddle-shaped stems, which add a welcome twist on texture in the garden. Low-growing succulents like golden barrel cacti (*Echinopsis grusonii*), gopher plant (*Euphorbia rigida*), and Moroccan mound (*Euphorbia resinifera*) can be planted in groups and used much like groundcovers.

You can create a landscape filled with only succulents or scatter them among trees, shrubs, and groundcovers. There is lovely texture contrast when succulents' unique shapes are viewed with the softer forms of leafy plants or ornamental grasses. Boulders are also the ideal partner for cacti and other succulents. These water-saving plants are an extremely versatile plant from a design standpoint. You can use them for contemporary gardens with straight lines and rows, but they work equally well in a more natural arrangement.

SHADOWS IN THE GARDEN

One design element often overlooked in the garden is the shadows that plants can reflect, adding another layer of interest. Not surprisingly, succulents' curved and spiky shapes create nice shadows on vertical surfaces when the sun shines on them. A bare wall can be transformed by placing plants when oriented in an east- or west-facing direction.

Artichoke agave (*Agave parryi v. truncata*), golden barrel cacti (*Echinocactus grusonii*), lady's slipper (*Euphorbia lomelii*), and blue yucca (*Yucca rigida*) create an attractive, low-water-use entry design.

A pair of boulders adds height and texture contrast alongside the base of a saguaro cactus (*Carnegiea gigantea*), artichoke agave (*Agave parryi v. truncata*), and lady's slipper (*Euphorbia lomelii*).

THE HIDDEN COSTS OF BOULDERS

While the cost of boulders/large rocks is based on their weight, big ones can be hard to maneuver and place and may require a crane or other heavy equipment to move. For small to medium-scale gardens, group two smaller rocks together instead of one large one to create the appearance of a bigger stone. They are easier to move and have the same design impact.

BOULDER PLACEMENT

We include boulders (large rocks) in this section because they are a popular element for dry climate garden design and make up much of the natural landscape surrounding us. Rocks can be used to create walls for raised beds or retaining walls.

They also serve as a great visual anchor for plants. Pair a boulder next to a plant, and it looks better due to the mixture of heights and differences in texture. They are a great way to add height or be placed alongside raised areas or swales in a relatively flat landscape. You can find boulders available where you buy your inorganic rock mulch. When choosing boulders for your garden, the size of the each one should be in scale with the size of your space. For example, if you have a smaller garden, select a boulder that is perhaps 2 feet (61 cm) in width and 16 inches (41 cm) tall—larger landscapes will need larger rocks. To create a more natural appearance, once you have your boulder in the desired position, bury the bottom one-third.

This backyard garden is a feast for the eyes with flowering groundcovers, shrubs, and vines combined with succulents.

The vibrant blooms of bougainvillea (*Bougainvillea spp.*), blue palo verde (*Parkinsonia florida*), and chaparral sage (*Salvia clevelandii*) provide the backbone colors of this shared community area.

HOW TO USE COLOR AND TEXTURE

The color palette in arid climates is filled with soft shades of green with undertones of gray or blue, which are characteristic of plants that have adaptations to limit water loss through the leaves. In addition, splashes of darker greens are also present with plants that look like they need a lot of water but are surprisingly drought tolerant. The unique shapes of cacti and other succulents add interest and artistry, while flowering plants add vibrancy to the landscape from soft shades of white to deep red. In short, there is no limit to the potential for color and texture in dry climates.

COLOR

People naturally gravitate toward color, and the landscape is an ideal place to incorporate it to up the curb appeal around your home. When people think of dry climate gardens, "colorful" isn't usually a term that comes to mind. However, many colorful plants make excellent choices for arid climates, with many that benefit pollinators. If you enjoy many colors, your initial inclination may be to add many plants in different colors. But, this can visually lead to a "busy" or messy-looking garden. Instead, we need to learn to use color most effectively and not create visual overload.

First, begin by choosing two to three primary colors to feature in your landscape, in addition to green. You will want to choose flowering plants that represent these colors. For example, you may have large shrubs that have purple blooms, two types of small shrubs with orange flowers, and then a yellow-blooming groundcover. It is okay to add a fourth or fifth color but use it on a smaller scale. The colors you select should be based on which ones you like best and the kinds of plants available in those shades. I recommend using at least one color that you have inside of your home. For example, if you have orange elements inside, choose a plant with that color to help tie both spaces together visually.

Warm colors, red, orange, and yellow, are used to grab your attention and can make an ample garden space appear smaller, which works nicely if you want to create a cozy atmosphere in the garden. On the other side, cool shades such as blue and purple are calming and can create the illusion of a bigger space. In my garden, I utilize both warm and cool colors. On a hot summer's day, I particularly enjoy the

WHAT COLORS GO WITH EACH OTHER?

To help you choose the colors in your garden, it is helpful to use the color wheel. To create a dramatic effect, select colors on opposite sides of the wheel, or for a softer blending, utilize plants whose colors occur next to each other on the wheel. When pairing different plants together for their colorful flowers, keep in mind that they may not bloom simultaneously, so note when they flower.

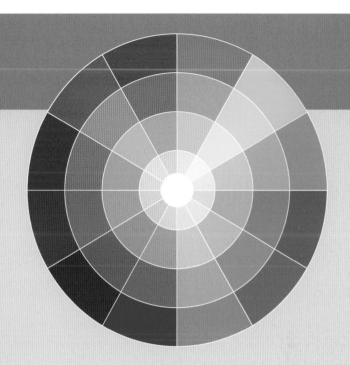

purple shades of my trailing lantana (*Lantana montevidensis*), Texas sage shrubs (*Leucophyllum spp.*), and pink trumpet vine (*Podranea ricasoliana*), which create a visually cooling effect even when temps are over 100°F (38°C). The orange blooms of my 'Sparky' tecoma (*Tecoma × 'Sparky'*) shrubs and yellow angelita daisy (*Tetraneuris acaulis*) create eye-catching interest to passersby in my front yard.

To make the most of color in the landscape, we need to do several things: First, we want to group the same plants together to maximize their presence and color impact. A single plant (unless it is large) can often get swallowed up by the rest of the landscape. Second, choosing the best locations to focus on placing colorful plants is essential. These areas may include the areas along the corners of the landscape (next to the driveway near the sidewalk/street), near the front entry path, underneath the front window, and any side walls that face the road. You do not need to add color in all of these locations, but if you are like me and love color, these are the spots to add it. For year-round interest, select plants with overlapping bloom seasons, and finally, repeat your favorite colorful plant throughout the landscape to help tie it together visually.

In the backyard, color is best used against a fence or wall across from a window, next to a pool, and near the patio or seating area. Large colorful shrubs are helpful for use against a fence or a vine if the space is narrow, as may be the case with a side yard. Low-litter plants are the best choice near pools and water features. Colorful containers filled with succulents are a great way to add a color splash without creating a mess. Flowering annuals are always a dependable choice for color and are best used near a patio where they are best viewed close up.

Firecracker penstemon (*Penstemon eatonii*) and angelita daisy (*Tetraneuris acaulis*) are grouped to provide colorful winter interest alongside a driveway in a central Arizona garden. The same plants are repeated in other high-profile areas.

Colorful cushions, throw pillows, and a painting add year-round interest to this outdoor space.

Plants aren't the only way to introduce color to the landscape—other elements such as outdoor furniture, garden art, painted walls, and pots can also add a welcome splash of color. Throw pillows or cushions in a vibrant hue create year-long interest and is an easy way to spruce up the outside of your home. Paint your fence or wall to add a colorful background or hang garden art. My favorite way to add color to the dry climate garden is to incorporate colorful pots planted with low-maintenance succulents. Be selective in where you decide to add color in your front and backyard for maximum impact. In the areas between your concentrated color plants, add other plants with attractive foliage that serve as a visual filler.

TEXTURE

Texture within the garden is the visual size and shape of the plant and its leaves and is often described as bold, soft (fine), or coarse. Cacti and other succulents add bold texture, flowering shrubs and ornamental grasses offer soft texture, and trees are an example of coarse texture. An easy way to add interest throughout the year in your garden is to combine plants with different textures. Rock is another way to introduce texture to your outdoor space.

The fine texture of coral fountain (*Russelia equisetiformis*) contrasts beautifully with the unique blocky shape of the totem pole cactus (*Pachycereus schottii f. monstrosus*).

MINIMIZING URBAN HEAT ISLANDS

Weather patterns show that our outdoor spaces are only getting hotter, whether you are talking about your region or your home's landscape areas. Thankfully, plants can help decrease temperatures outdoors when used in strategic locations. Trees can be placed to provide shade in places around your home that get blasted by the afternoon sun and in south-facing exposures that get sun all day. The walls of your home and block walls around a backyard increase the heat as they absorb the sun's rays and re-radiate the heat back out. Tall shrubs and vines can be grown against these structures to keep the sun from heating them. Landscape rock can also re-radiate heat. Groupings of groundcovers prevent this, which is particularly useful in spots near driveways, sidewalks, and streets that absorb heat and release it back out.

FIREWISE LANDSCAPING

Wildfire threat in arid regions is prevalent, particularly when native vegetation dries out in the summer months. If you live in an area where wildfire is a factor, strategies to reduce the threat of fire to your home can be integrated into your landscape design. Plants can contribute to the danger of wildfire on your property. Thankfully, there are strategies that you can incorporate to reduce this danger, from plant placement to the type of plant you choose.

STEPS TO REDUCE FIRE RISK

- Create a defensible (plant-free) zone—don't add plants up against your house.
- Tree branches should be kept at least 10 feet (3 m) from any structure.
- Use smaller plants—2 feet (62 cm) tall or more petite—near your home, spacing wider apart. Larger plants can be added further outward.
- Avoid planting large masses of plants together.
- Prune plants to get rid of old or dead growth. Clean up all plant debris.
- Keep plants well-watered near the home and avoid letting them dry out.
- Use succulents or native and arid-adapted plants that have smaller leaves. Deciduous plants are also a good choice.
- Avoid using plants with a higher resin level, such as juniper, fir, pine trees, and palms, which are more flammable than many other plants.

You can get additional information from your local forestry or fire departments for your particular region, including recommended plant lists.

Young hibiscus shrubs planted in a west-facing exposure near a wall. The intense afternoon sun and radiated heat from the wall is too much for these tropical natives. A plant label that specifies "full sun" doesn't mean that plants can tolerate areas where reflected heat is present. Look for plants that can tolerate hot, reflected sun instead.

Several different types of shrubs are planted too closely and excessively pruned. There's not enough room to grow to their natural shape and size. Choose fewer shrubs and allow adequate spacing to grow to their natural shape and size.

Groundcover form of rosemary (*Rosmarinus officinalis* 'Prostratus') that grows 3 feet (91 cm) wide is misused in a planting space that is 4 inches (10 cm) wide, resulting in incorrect pruning and an awkward appearance. Additionally, the irrigation is too close to the foundation of the house. An alternative is to fill the space with decorative small-sized river rock and add an upright container with a succulent, which needs little water.

Desert spoon (*Dasylirion wheeleri*) and red yucca (*Hesperaloe parviflora*) butchered by incorrect pruning practices. Their leaves should never be "trimmed" on the top. Instead, prune outer leaves back to the base if needed. Learn how to maintain your plants you have the right way or ensure your landscaper does.

A pair of drip emitters continue to water nonexistent plants. Smaller plants such as groundcovers don't live as long as trees and shrubs, and often aren't replaced when they are removed. When you remove old plants, add new ones or plug unused drip emitters so you aren't wasting water.

A modern home's front landscape is filled with the spiky shapes of blue agave (*Agave americana*), large golden barrel cacti (*Echinocactus grusonii*), and flowering Bells of Fire™ Tecoma (*Tecoma x Bells of Fire*™) shrubs.

COMMON DESIGN MISTAKES AND HOW TO AVOID THEM

Often, we may not realize the mistakes we make in the landscape. So, I am sharing photos of problems to help you recognize these mistakes to help you avoid or correct them. After seeing them, I promise you will begin seeing them everywhere—in your neighborhood and driving around town.

SAMPLE LANDSCAPE DESIGN LAYOUTS

A well-designed landscape should complement your home while allowing you to enjoy spending time within it. It is helpful to think of your garden as an extension of the inside of your home—it is meant to be functional yet attractive at the same time. However, it isn't always easy to know where to start when configuring the design of your outdoor space. To help you, I have created five designs based on different criteria from design style, ease of care, size of the yard, and the types of hardscape elements within it.

The goal of these sample designs is to inspire you that you can fully customize them to fit your style and specific needs based on your landscape. Feel free to switch out the plants for ones you prefer or utilize some of the hardscape elements as a starting point in designing your own. If working with a landscape designer or contractor, these designs can show them what you want.

One word of caution: Some landscape professionals tend to overplant and not allow for the mature growth of plants, so do your homework and make sure plants have enough room to grow—it will save you money, maintenance, and water. You'll have a nice-looking garden as a result.

Ultimately, these plans are yours to use to create a beautiful, functional outdoor space that you will enjoy for years to come!

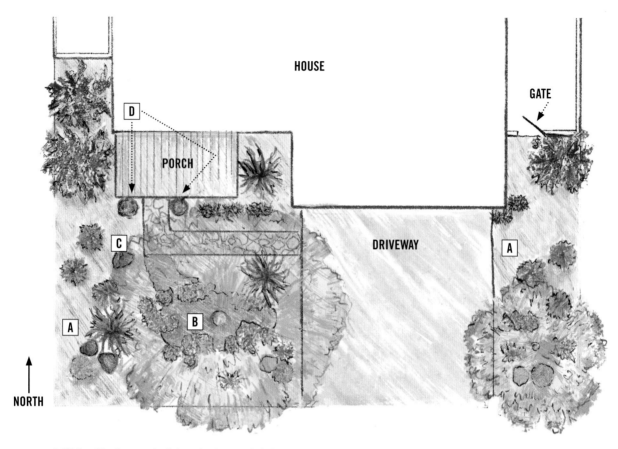

A: ½" (1 cm) Landscape rock   **B:** Larger landscape rock (6–9" rip-rap rock)   **C:** Boulders   **D:** Bright blue containers filled with trailing succulents

---

PLANT LIST

Artichoke Agave *Agave parryi v. truncata*
Octopus Agave *Agave vilmoriniana*
'Blue Elf' Aloe *Aloe* x 'Blue Elf'
Desert Willow Tree *Chilopsis linearis*
Golden Barrel *Echinocactus grusonii*
Blue Bells™ Emu *Eremophila hygrophana*
Purple Trailing Lantana *Lantana montevidensis* 'Purple'
'Desert Museum' Palo Verde *Parkinsonia* hybrid 'Desert Museum'

Elephant's Food *Portulacaria afra*
'Sparky' Tecoma *Tecoma stans* 'Sparky'
Angelita Daisy *Tetraneuris acaulis*

OTHER MATERIALS
½" (1 cm) landscape rock
Larger landscape rock (6–9" [15–23 cm] riprap rock) in swale (shallow depression 8–12" [20–30 cm] below existing grade) to encourage rain runoff to pool around the tree.

Boulders (2–3 feet [61–91 cm] wide, 16–20" [40–51 cm] tall, bottom ⅓ buried for a natural appearance)
Bright blue containers filled with trailing succulents

FRONT YARD CURB APPEAL

If I had to pick my favorite garden style, it would be filled with blooming shrubs and groundcovers coupled with cacti and other succulents. Plants that bloom at different times of year fill my front landscape, so there is something guaranteed to be blooming in my Phoenix-area garden. I love color and, like many people, gravitate naturally toward it. Thankfully, many colorful plants thrive in my hot, dry climate.

One of the strategies I recommend to up the curb appeal and interest of landscapes of my clients and students is to intersperse succulent plants among groundcovers and shrubs. This pairing does something special in the way they make each other more attractive. In addition, the unique outlines of succulents serve to highlight the softer shapes and textures of nearby flowering plants.

In this design, curb appeal is the focus, as it is with most front yards. The main focal point is provided by two desert trees, which provide filtered shade while still allowing enough sunlight to filter through for the trailing lantana to bloom. A gentle swale filled with larger-sized landscape rock surrounds the palo verde tree, directing rainfall from the house that helps to water the tree. 'Sparky' tecoma (*Tecoma × 'Sparky'*) shrubs offer a taller accent and help frame the residence's front visually. Low-maintenance elephant's food succulents are planted in bright blue containers that help to create an inviting entry. Boulders are spread out through the landscape, adding welcome height and texture to this design.

This xeriscaped front yard is attractive and offers year-round interest, and it is relatively low maintenance, only requiring pruning once or twice a year. So, as you can see, beauty in the garden doesn't have to take a lot of time and effort to maintain.

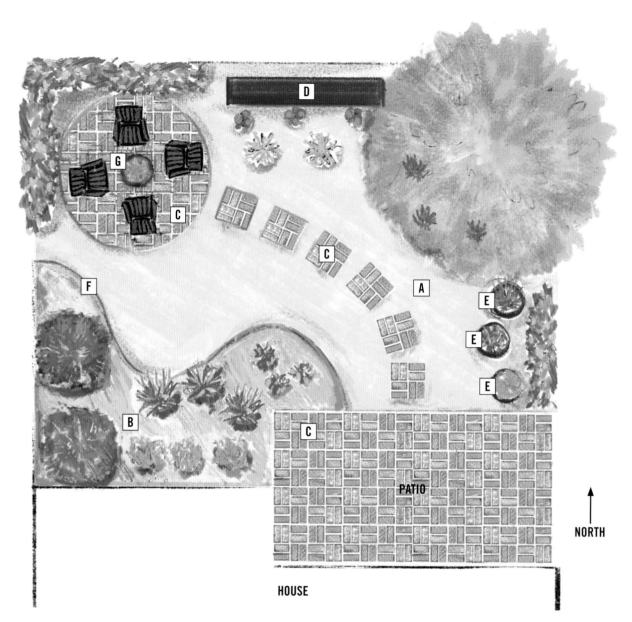

A: ¼" (9 mm) decomposed granite   **B:** 3" (8 cm) river rock   **C:** Gray pavers   **D:** 5-foot (2 m) tall painted low wall
E: Various potted flowering shrubs   **F:** Metal edging   **G:** Chairs and firepit

PLANT LIST
'Blue Elf' Aloe *Aloe* × 'Blue Elf'
Gopher Plant *Euphorbia rigida*
Firecracker Bush *Hamelia patens*
Mexican Honeysuckle *Justicia spicigera*
Totem Pole Cactus *Pachycereus
   schottii f. monstrosus*

Passion Vine *Passiflora incarnata*
Mastic Tree *Pistacia lentiscus*
Mealy Cup Sage *Salvia farinaceae*
Shrubby Germander *Teucrium
   fruiticans* 'Azurea'
Angelita Daisy *Tetraneuris (Hymenoxys)* acaulis

OTHER MATERIALS
¼" (9 mm) decomposed granite
3" (8 cm) river rock
Gray pavers
5-foot (2 m) tall painted low wall
Various potted flowering shrubs
Metal edging
Chairs and firepit

SMALL SPACE POLLINATOR GARDEN

Modest-sized gardens may not have the size of larger spaces, but they do present unique opportunities to create beauty and interest—just on a smaller scale. The key is to utilize the available area efficiently. This backyard design combines eye-catching design and creates a haven for pollinators using plants that attract bees, butterflies, and hummingbirds.

To make the most of this area, hardscape elements such as a patio and a firepit with step stones connecting them allow this space to be used throughout the year. In this design, gray pavers are used, but concrete would create a similar effect. The patio provides welcome shade for a BBQ grill, with room for a table and chairs or comfortable outdoor chairs and a coffee table. Ceiling fans or misters can be added to help cool the patio area during the hot summer months to allow for year-round enjoyment of the outdoors. The last hardscape element is a 5-foot-tall (2 m) stucco block wall painted a deep shade of purple that adds vibrant color and highlights the distinctive shape of the columnar totem pole (*Pachycereus schottii f. monstrosus*) cacti. The main landscape area is covered in ¼-inch (6 mm) decomposed granite to make walking easy while providing a smooth surface.

Plants are used to complement and frame the hardscape elements. A mastic (*Pistacia lentiscus*) anchors the corner where aloes enjoy the shade. The rest of the plants in this space are used for their beauty and ability to attract pollinators. Passion vine (*Passiflora foetida*) adds a nice vertical element and its purple flowers are simply lovely. Metal edging is used to create a border for a collection of flowering pollinator plants surrounded by 3-inch (8 cm) river rock that adds a decorative element. A trio of potted shrubs in colorful containers complete this outdoor space and connect to the irrigation system, making their care easy.

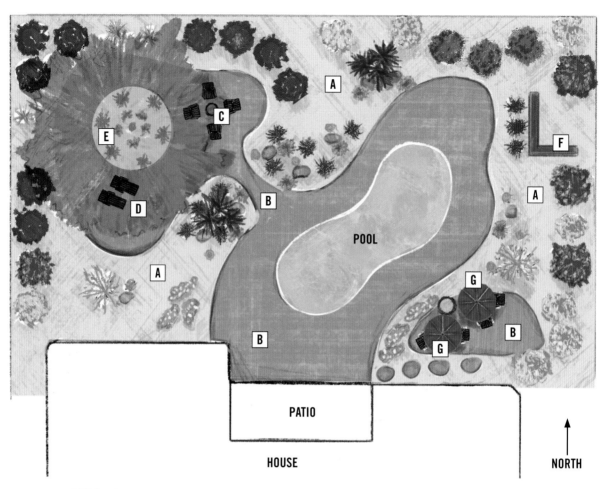

A: ¼" (6 mm) decomposed granite   **B:** Flagstone patio   **C:** Table   **D:** Lounge chairs   **E:** 3" (8 cm) river rock for around the tree
F: Low wall blocking the view of the pool equipment   **G:** Chairs and firepit with umbrellas

PLANT LIST
Artichoke Agave *Agave parryi v. truncata*
Cape Aloe *Aloe ferox*
Foxtail Asparagus Fern *Asparagus densiflorus* 'Myers'
Chocolate Flower *Berlandiera lyrata*
Red Bird-of-Paradise, Pride of Barbados *Caesalpinia pulcherrima*
Mediterranean Fan Palm *Chamaerops humilis*
Bush Morning Glory *Convolvulus cneorum*

Hop Bush *Dodonaea viscosa*
Ocotillo *Fouquieria splendens*
'Lynn's Legacy' or 'Rio Bravo' Texas Sage *Leucophyllum langmaniae*
Pink Muhly Grass *Muhlenbergia capillaris*
Fruitless Olive *Olea europaea* 'Swan Hill'
'Crimson Flare' Tecoma *Tecoma* 'Crimson Flare'
Purple Heart *Tradescantia pallida*

OTHER MATERIALS
¼" (6 mm) decomposed granite
3" (8 cm) river rock for around the tree
Flagstone patio
Furnishings

LARGE BACKYARD WITH POOL

The goal of many homeowners in hot, dry climates is to enjoy outdoor living all year long. As a result, swimming pools are a common presence in the backyard. In some landscape settings, the yard is small, and the pool takes up most of the room, while large backyards may have a sizeable landscape-able space. No matter the size of your backyard space or pool, the plant selection should be based on lower-litter plants.

This backyard has three main components—the pool, a sizeable seedless olive tree, and flagstone seating areas. Flagstone is the decking choice that is a classic option that won't go out of style and lends itself to landscapes with curves. Decomposed granite is used throughout the backyard except for the area around the tree where 3-inch (8 cm) river rock is placed. The olive tree is large, evergreen, and relatively low litter, and its shade beckons people to sit in one of the seating areas. A smaller seating area with a fire pit sits on the other side of the pool with colorful umbrellas for shade.

With larger spaces, it's essential to tie the entire area together visually, or it may look like unconnected blocks of space. The easiest way to achieve this is to repeat certain plants throughout the landscape. In this backyard design, several shrubs appear on both sides of the yard and along the back wall, which helps to create that continuity.

Tall shrubs cover the bareness of the block walls while reducing reflected heat in summer. The unique shapes of succulents create interest with large rocks to visually anchor their base. The ornamental grasses provide a soft, flowing texture between the pool and tree, while shade-loving plants thrive under the tree's shade. Shorter shrubs and groundcovers are placed to create lower interest throughout the year.

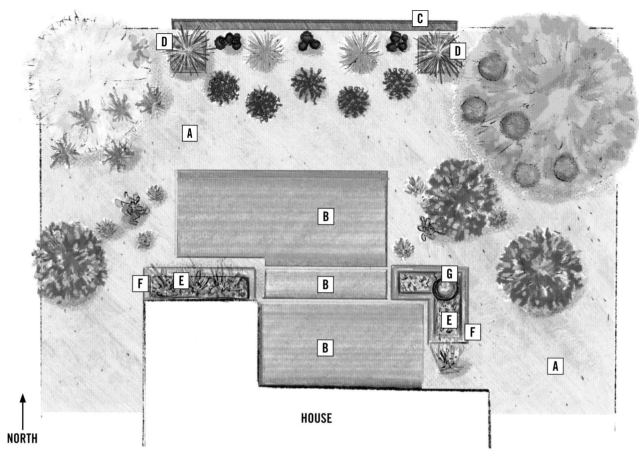

NORTH

A: ¼" (6 mm) decomposed granite   **B:** Concrete patio and matching steps   **C:** 5-foot (2 m) painted wall   **D:** Plinth with container on top
E and F: 3- to 6-inch (8–15 cm) rip-rap rock enclosed with concrete border   **G:** Pot fountain

PLANT LIST
Mulga *Acacia aneura*
Twin Flower Agave *Agave geminiflora*
Desert Marigold *Baileya multiradiata*
Cascalote *Caesalpinia cacalaco*
Desert Spoon (Sotol) *Dasylirion wheeleri*
Texas Mountain Laurel *Dermatophyllum
    secundiflorum*

Golden Barrel *Echinocactus grusonii*
'Brakelights' Red Yucca *Hesperaloe
    parviflora* 'Brakelights'
Hardy Spineless Prickly Pear *Opuntia
    cacanapa* 'Ellisiana'
Mexican Fence Post *Pachycereus marginatus*
Lady's Slipper *Pedilanthus macrocarpus*
Elephant's Food *Portulacaria afra*

OTHER MATERIALS
¼" (6 mm) decomposed granite
Concrete patio and matching steps
5-foot (2 m) painted wall
18-inch (46 cm) plinth with container on top
3- to 6-inch (8–15 cm) riprap rock
Pot fountain

FUSS-FREE LOW-WATER-USE GARDEN

An attractive outdoor space that requires little maintenance and uses less water is possible, as this design illustrates. Hardscape elements play a focal point with the concrete patio being extended out into the garden. The concrete areas can be used to provide for outdoor living by adding a table for dining, a fire pit, seating areas, and more. A colorful pot fountain adds interest and the welcome sound of water alongside the patio. The main view from the house rests on an orange stucco wall that serves as a vibrant backdrop for columnar cacti, fan-shaped succulents, and potted agave.

The primary plant choices are cacti and other succulents, which are very low maintenance, yet add beauty with their unique shapes. Another benefit of using succulents is that they don't need much supplemental water. Using succulents with bold shapes and textures helps provide some softness at their base with the softer outlines of groundcovers. Flowering desert marigold (*Baileya multiradiata*) adds color and delicate texture and does well on the same watering schedule as succulents. Many succulents that provide seasonal blooms are incorporated into the landscape, adding another splash of color. Utilizing plants to provide for varying heights within the landscape creates a cohesive and pleasing design. Two arid-adapted trees help to provide needed height while anchoring the bare corners. Large shrubs that have been trained into tree shapes serve as a transition between the larger trees and lower growing plants.

The maintenance in this garden is low due to the plant selection. While trees can be messy and high maintenance, the types chosen for this space are relatively lower litter and need pruning once a year or less. Functional yet beautiful outdoor spaces don't have to be much work to care for.

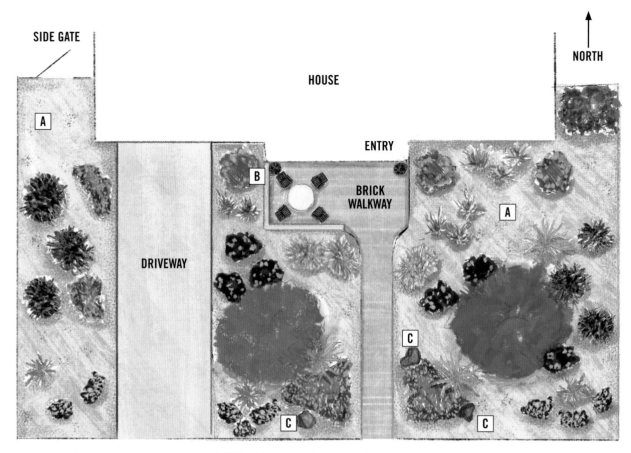

A: ½" (1 cm) landscape rock **B:** Fence **C:** Boulders

PLANT LIST
Murphy Agave *Agave murpheyi*
Aloe Vera *Aloe barbadensis*
Damianita *Chrysactinia mexicana*
Skyflower *Duranta erecta*
Valentine Bush *Eremophila
   maculata* 'Valentine'

'Red Push' Pistache Tree (*Pistacia* x 'Red Push')
Red Yucca *Hesperaloe parviflora*
Moss Verbena *Glandularia pulchella*
Gold Lantana *Lantana* × 'New Gold'
Firecracker Penstemon *Penstemon eatonii*
Coral Fountain (Firecracker Bush)
   Russelia equisetiformis
Yellow Bells *Tecoma stans*

OTHER MATERIALS
½" (1 cm) landscape rock
Fence
Brick-paved area with furnishings
Boulders (2–3 feet [61–91 cm] wide,
   16–20" [41–51 cm] tall, buried ⅓ deep)
Brick walkway

YEAR-ROUND COLOR GARDEN

Drive through any neighborhood, and you are likely to see front yards covered in thirsty grass or artificial lawns. However, this isn't the only option for the front of your home. You can create a vibrant garden that flourishes in a hot, dry climate while using a fraction of water that a lawn requires. Many homeowners are ripping out their grass, looking for other options, and adding blooming plants that flower at different times of the year, which can ensure colorful year-round interest.

People like lawns because of the vibrant green color; however, it isn't the only way to add lush green to the front yard. In this design, trees, shrubs, and groundcovers with deep green foliage provide an excellent (non-walkable) substitute for grass. Plants were also selected for flower or foliage color that they added. One error I see people making is that all their plants bloom at the same time of year, leaving their garden rather colorless at other times. In dry climate regions that experience mild winters, you can have plants that bloom in the cool season while others flower at warmer times of the year. In short, we want to plan for overlapping bloom seasons when possible.

A front courtyard area was added to increase the outdoor space for use. It is surrounded by 3-foot-tall (91 cm) white wooden plank fencing to create separation from the rest of the front yard. Brick is a classic choice for a hard surface and adds nice color.

Other elements can be introduced that will add vibrant interest, even in the cold of winter, by using colorful containers. The texture is another way to ensure year-round interest by using large rocks and succulents among the shrubs and groundcovers.

Create a mini-pollinator garden in a pot, like this one with tropical milkweed (*Asclepias curassavica*) and scabiosa (*Scabiosa spp.*).

DRY CLIMATE CONTAINER GARDENING

Growing plants in containers is a great way to increase the room to grow plants—even if you don't have a front or backyard. It also allows you to create islands of inviting color and interest around the outside of your home. Containers aren't just vessels for plants; they can also add a decorative element throughout the garden or against your house. We can do it all from flowering annuals, shrubs, cacti, and other succulents! While countless plants do well in pots in dry climates, there are some challenges that we need to plan for and mitigate to ensure that our container plants thrive while adding beauty to our outdoor spaces.

LOCATION

We need to match the plants we want to grow to their preferred exposure—see the How the Sun Affects Plants section in Chapter 1. However, the stress that the sun and heat create in pots is amplified as the roots aren't in the cool ground and dry out faster. Select an area that receives at least five to six hours of sun for most plants. Avoid placing pots against west-facing walls, which is a very stressful location—moving the pot a few feet away will help. If you have a shady area, it is hard to find plants that will flower and add color. Instead, add a pot in your favorite shade of blue or another vibrant color, and plant a shade-loving plant in it.

CHOOSE THE RIGHT CONTAINER

There are many options for containers, from the size, shape, color, texture, and material they are made from. A pot's most important thing is to have a hole for drainage at the bottom. Secondly, the size of the pot is a big deal as the hot air temperatures in summer can literally "cook" the roots of plants near the sides of the container, and the soil dries out fast. This is why plants in smaller containers struggle to survive in summer. In short, bigger is better for containers as it provides some insulation for the roots from the heat, and it holds onto more water and dries out more slowly. A good container size for annuals and other plants up to 2 feet (61 cm) tall is 16 inches (41 cm) wide and tall. The larger a plant is, the bigger pot it needs.

TYPE OF CONTAINER	PROS	CONS
Glazed Ceramic	Most colorful and decorative options, last a long time	Hard to move and can be expensive
Concrete	Available in many shapes and earth-tone tints, resistant to hot weather	Very heavy and difficult to move
Plastic	Inexpensive, easy to move, some mimic real stone or concrete in appearance	Can become brittle and crack from sun exposure
Terra-Cotta	Does well in hot weather and comes in countless sizes and shapes	The porous clay of terra-cotta dries out fast, requiring frequent watering
Wood	Inexpensive choice, doesn't conduct heat	Decays after a few years unless sealed regularly

AVOIDING UNSIGHTLY POT STAINS

Containers can stain your patio when water runs off, and if you use a saucer, that is an invitation for mosquitos. To avoid this problem, place pots in cut-out areas around your patio where they will rest on the dirt. Surround them with decorative inorganic rock mulch. If you are designing your backyard, include these cutouts in the plan. A bonus is that you can often connect container plants to irrigation lines.

Type of Container

Pots are made from various materials, with some being more suitable for dry climates.

From a decorative standpoint, I favor using glazed ceramic pots or plastic pots that look like natural stone—both perform well in dry climates. You can transform a dull, plastic pot with color spray paint.

The author planting a combination of annual flowers and perennial plants in a container that includes calendula, celosia, and rock verbena.

SOIL MIX

When walking into your favorite nursery or big box store, you may find yourself faced with numerous different bags of potting soil. It can be confusing when faced with many choices—some with added fertilizer, others with water-holding gels, and some with neither. The most crucial factor in choosing a soil mix for your pots is to look for one formulated for containers, which will be clearly stated on the label. These soil mixes are created for container gardening and don't hold onto too much water, which is essential as plants don't like soggy soil. Cacti and other succulents require a well-drained soil mix, so buy a planting mix that is specially formulated for these types of plants. You can save money and make your own soil mix for succulents by mixing three-parts regular potting soil, two-parts coarse sand (not fine), and one-part perlite.

Cover the drainage hole with a single rock to prevent soil from falling through the hole. You can also add a piece of wire mesh, which will do the same. There is a common myth that adding a layer of small rocks or gravel at the bottom of a pot will improve drainage, but the opposite is true—it can lead to water-logged soils. When buying multiple bags to fill a large pot, soil mixes can get expensive. You can fill the bottom quarter to one-third of each container with empty plastic soda or water bottles for shallow-rooted plants such as annuals, cacti, and succulents. This saves you money on soil mixes and makes the pot lighter and easier to move around.

WATERING CONTAINERS

I'm most often asked about container gardening: "How often do I need to water?" It depends on several factors, including the type of plant, outside temperature, whether the pot is in full

WATERING SUCCULENTS: LESS IS BEST

Succulents don't need to be watered often. They struggle and will even die when overwatered. This is a typical case of killing plants with kindness (in this case, too much water). Allow the soil to dry between watering cycles in warmer months and water every three to four weeks in winter. Cacti need less water than smooth succulents. Before watering, use a screwdriver to poke holes in the dry soil to help the water to penetrate the root ball. Look for signs of wrinkling leaves, which is a sign of water stress.

sun or shade, wind conditions, and the material the pot is made of. Now, you can buy a soil moisture meter, but they aren't always reliable, and there is a much simpler way to determine when to water container plants that I call the "finger-test." Take your finger, stick it down into the soil at the recommended depth below, then pull it out. If there is some soil clinging to your finger, it is still moist and doesn't need water. However, if your finger comes out relatively clean, it's time to water. The hotter and drier it is, you will need to water more often, while cooler temperatures mean less watering is required.

- Annuals—water when the top inch of soil is barely moist
- Perennials and shrubs—water when top 3 inches (8 cm) is almost dry

Water your pots until you see water coming out of the bottom, as it's important to wet the entire root ball, which helps to flush out accumulated salts and makes the interval between watering longer. You can water containers by hand with a hose or watering can but consider connecting them to a drip irrigation system to automate watering if you have many containers.

FERTILIZING CONTAINER PLANTS

Anything that grows in a container needs to be fertilized—even cacti and other succulents. The reason is that plants don't have access to nutrients in the soil, so we must provide them. There are a variety of fertilizers that work well for plants grown in containers, from slow-release fertilizer granules, fertilizer spikes, and liquid fertilizer. You can find all these types in organic and synthetic forms. Select a fertilizer with equal amounts of nitrogen, phosphorus, and potassium. I prefer to use slow-release granular fertilizer to feed my plants, and I don't have to apply it often. For flowering annuals, liquid fertilizers with a higher ratio of phosphorus can be used monthly to promote blooming in addition to a slow-release fertilizer. Succulents don't need as much care as other container plants, especially when it comes to fertilizer and water. Apply a slow-release granular fertilizer to potted cacti and different succulents once a year in spring.

Apply fertilizer at the beginning of spring, following package directions. Except for flowering annuals, most potted plants don't need fertilizer through the winter months as this can promote frost-tender growth. In general, the last application of fertilizer should occur six to eight weeks before the first frost date for your region.

Container filled with succulent plants, firesticks (*Euphorbia tirucallii* 'Sticks on Fire'), and elephant's food (*Portulacaria afra*).

Arranging flowering annuals can be fun. One popular design tool is called "thriller, filler, and spiller." In the author's garden, she has used this technique for her raised flowering bed. The tall center plant, Gomphrena 'Fireworks', is the "thriller"; the medium-size plant, white vinca, is the "filler"; and the outer plants, purple gomphrena, are "spillers" that trail over the edge of the container.

The unique shapes of succulents are amplified against a colorful blue wall at the Tucson's Botanical Gardens in Tucson, Arizona.

TYPES OF CONTAINER PLANTS

You can grow just about any type of plant in pots. Think beyond flowering annuals when it comes to container plants. My favorite color choice is to use flowering shrubs as I get the welcome color of blooms without the fuss that flowering annuals require. Consider growing succulents in containers if you want to add pots but don't want to fuss over plants.

Annuals

Annuals in vibrant shades are the most popular pot filler, and they grow surprisingly well in our dry climate. The different shapes and colors allow you to create a living bouquet to decorate the areas around your home. Place them in high-visibility areas along your front walkway, in front of patio pillars, or around your pool—just be sure that they will receive at least six hours of sun, or you may not get blooms. Cool-season favorites are planted in fall in mild-winter regions, including alyssum, geraniums, lobelia, petunias, snapdragons, and violas. Wait until daytime temperatures are 85°F (29°C) or lower before planting in fall, or they will struggle. Warm-season annuals that do well in hot summers are angelonia, celosia, gomphrena, salvia, and zinnias, which should be planted in mid to late spring before temperatures get above 90°F (32°C) to allow their roots to grow before the heat of summer arrives.

In the author's garden, colorful pots filled with flowering shrubs such as Blue Bells™ Emu (*Eremophila hygrophana*), Mexican honeysuckle (*Justicia spicigera*), and mealy cup sage (*Salvia farinaceae*). Coral fountain (*Russelia equisetiformis*), planted in the ground, adds a nice splash of orange color.

The aromatic blooms of chaparral sage (*Salvia clevelandii*) decorate a front yard in spring.

Succulents/Cacti

One type of container gardening that we excel at in arid climates is using succulents, including cacti. They do well because they like it when it is dry outside and dislike soggy soil, so they are tailor-made for the natural conditions in our gardens. You can create eye-catching succulent arrangements that showcase their unique shapes. Succulents are also incredibly versatile and can be used in areas with no nearby water source as you will only need to water them sporadically. I like to add a decorative layer of small river rock around succulents to create a sophisticated look.

Shrubs

After years of growing high-maintenance annuals in pots, I decided that I wanted something easier to care for while still adding beauty to my containers. So, I began planting flowering shrubs in pots, and I've never looked back. My favorite spot in my back garden is a grouping of blue, purple, and yellow pots filled with colorful shrubs. Butterflies and hummingbirds are frequent visitors to their nectar-rich flowers and are enjoyable to observe. Shrubs need a good-sized pot at least 2 feet tall (61 cm) and 16 inches (41 cm) wide—bigger is even better. Shrubs will be smaller in containers than planted in the ground as their roots are restricted. Spring pruning for shrubs in pots involves cutting them back more severely than those in the ground to avoid outgrowing the container. I encourage you to try growing your favorite shrubs in pots for an attractive yet lower-maintenance pot filler.

There are countless options for container gardening in a dry climate, and I encourage you to experiment with different combinations and styles. You will be successful when choosing larger pots and the right plant(s) and fertilizing and watering when dry.

CHAPTER 5

PLANT LIST

The key to success in gardening in a dry climate comes down to selecting the right plants. Most frustration in the garden is due to growing ill-adapted plants that struggle to survive. To help you avoid this frustration, I have created lists of my favorite trees, shrubs, groundcovers, grasses, palms, and cacti/succulents for this book. Of course, these are just a partial listing of suitable plants for dry regions. While all can handle warm to very hot summers, there are varying levels of cold hardiness with some plants adapted to sporadic freezing temperatures to others that can handle frigid winter conditions. Minimum cold tolerance is provided so you can select the right plants based on how cold your winters are. In addition to the individual plant profiles, you will find charts that showcase even more plants for specific situations, such as pollinator gardens, reflected heat, shade, etc.

For each plant, note the approximate size and recommended exposure, which will help to ensure that they will perform well in your garden. Unless specified, all plants do best in well-drained soil and don't require supplemental fertilizer. While many of these plants may be available at your local garden center or nursery, if they aren't, ask your nursery professional if they can order them for you. Or, see if your botanical garden has seasonal plant sales where you are likely to find a large number of plants perfect for your specific region.

Within the plant lists, common names are added along with botanical (scientific) names. Use botanical names to ensure you are getting the right plant. Botanical names may change from time to time, but when used along with common names, you'll be able to get the plant you want.

As you go through the plant list, write down your favorites and look for complementary pairings (plants that will look good together). Note when the bloom times are, particularly if you want to enjoy the flowers of two different plants simultaneously or if you want to ensure overlapping bloom seasons.

When choosing plants, don't be afraid to go with a smaller container, particularly if it's a moderate to fast-growing plant. Younger plants are easier to plant, grow faster initially, and are more adaptable than those in larger containers. This is a great way to save money and the difference in size in a year between planting a 1-gallon and a 5-gallon (5–23L) plant is negligible. However, if a plant grows slowly, you may want to go with the larger size specimen. Ultimately, the choice is up to you. For a large planting project, you may not be able to find all the plants you will need at one time. That's okay. Just get what you can and purchase the other plants at a different time of year once they become available.

TREE PROFILES

Acacia aneura

MULGA

Bloom Period: Spring, Summer

Mature Height x Spread: 18 × 16 ft (5.5 × 4.9 m)

Exposure: Full sun

Cold Tolerance: 15°F (-9°C)

Origin: Australia

Evergreen tree with silvery-gray foliage with narrow leaves (phyllodes). Mulga trees grow naturally in a large shrub shape but are often trained in tree form. They have a nice rounded canopy with tiny, golden yellow oblong flowers that appear in spring. Repeat bloom can occur in summer, especially in response to rain or increased humidity. This tree has a slow growth rate, is thornless, and produces seedpods. Mulga makes an excellent privacy screen and windbreak when grown in shrub form. Flowering groundcovers and succulents do well when planted underneath as the foliage creates filtered shade. This Acacia species is susceptible to overwatering, so water deeply and infrequently. Don't plant in a lawn.

Caesalpinia cacalaco

CASCALOTE

Bloom Period: Fall to Winter

Mature Height x Spread: 15 × 15 ft (4.6 × 4.6 m)

Exposure: Full sun, reflected heat

Cold Tolerance: 20°F (-7°C)

Origin: Mexico

Evergreen trees with dark-green foliage are made up of circular leaflets. This medium-sized tree has a lovely traditional tree shape and grows slowly. Yellow flowering spikes appear mid-fall and last into winter, adding welcome color to cool-season landscapes. Seedpods form once flowers fade. A notable characteristic is its large, rose-like thorns, so keep it away from pathways or sidewalks. Cascalote is an excellent option for reflected heat areas and can provide welcome afternoon shade. Because it is a reasonably low-litter tree, it is suitable for use near seating areas or to anchor a west-facing corner of your garden. Prune in spring, removing any remaining seedpods at the same time. 'Smoothie' is a thornless variety.

TREE PROFILES

Chilopsis linearis

DESERT WILLOW

Bloom Period: Spring, Summer

Mature Height x Spread: 25–30 × 25–30 ft (8–9 × 8–9 m)

Exposure: Full sun, reflected heat

Cold Tolerance: -10°F (-23°C)

Origin: Southwestern United States

Deciduous tree with bright green foliage that consists of narrow leaves. In mid-spring through summer, branches are covered with large pink flowers. There are multiple varieties available with flowers colors in varying shades of pink, from light to dark. Narrow seedpods remain on the tree once the flowers drop; however, newer varieties have minimal to no seedpods. The natural shape of desert willow is a large shrub, but they are often trained into traditional tree forms. This tree has a moderate to fast growth rate and is suitable for smaller garden spaces as it can be easily maintained at a smaller size—plant near a seating area to provide light shade or frame a view from a window. Prune to the desired shape in late winter and remove tree suckers throughout the summer as needed.

Cordia boissieri

TEXAS OLIVE, MEXICAN OLIVE

Bloom Period: Spring, Summer, Fall

Mature Height x Spread: 25 × 25 ft (8 × 8 m)

Exposure: Full sun, reflected heat

Cold Tolerance: 20°F (-7°C)

Origin: Texas, Mexico

The deep green leaves of Texas olive have a soft, leathery texture and provide lush color to the landscape throughout the year. Like many natives of the Southwest, its natural shape is shrublike but is often maintained in a traditional tree shape. Butterflies are attracted to the large white flowers that have a papery appearance and bloom off and on over a long period. The blooms are followed by a small green fruit that resembles an olive and is edible. The flowers and seeds cause seasonal litter, so don't use them near a pool or alongside a patio. Texas olive makes excellent privacy screens or a privacy hedge when allowed to remain in its natural shrub shape. This slow-growing tree/shrub makes a lovely smaller tree for a more compact landscape and can tolerate reflected heat. Prune to the desired shape in spring.

TREE PROFILES

Dermatophyllum secundiflorum
(syn. Sophora secundiflora)

TEXAS MOUNTAIN LAUREL

Bloom Period: Spring

Mature Height x Spread: 10–15 × 15 ft (4–5 × 5 m)

Exposure: Full sun

Cold Tolerance: 10°F (-12°C)

Origin: Mexico, Texas

The dark green foliage and small size of this tree make it a great foundation plant in smaller spaces such as a courtyard or patio. In spring, large clusters of purple flowers are produced with the delightful fragrance of grape candy. The flowering period lasts just a few weeks, and light tan seedpods soon follow. Within each pod is a toxic red seed, but seedpods can be cut off if desired. The natural growth habit of Texas mountain laurel is a large shrub and can be utilized as an evergreen screen. Alternatively, it can be trained into tree form. 'Silver Peso' is a variety with silvery-gray foliage. Prune after flowering to the desired shape but don't shear.

Ebanopsis ebano
(syn. Pithecellobium flexicaule)

TEXAS EBONY

Bloom Period: Summer

Mature Height x Spread: 25 × 20 ft (8 × 6 m)

Exposure: Full sun

Cold Tolerance: 15°F (-9°C)

Origin: Texas, Mexico

The evergreen foliage makes this tree a desirable addition for those who yearn for deep greens in their landscape. Clusters of tiny leaflets create a lush appearance, creating an oasis-like feel. The branches are pale gray and make a nice color contrast with the dark leaves. Texas ebony is a slow-growing tree, and cream-colored flowers appear from summer to fall, followed by large brown seedpods. As a result, this tree shouldn't be used near pools or areas that will require regular cleanup. The branches are covered in thorns, so Texas ebony is best used away from patios, pathways, or sidewalks. Use it as the focal tree in your front or backyard, where you can view its attractive foliage. Prune in spring to the desired shape.

TREE PROFILES

Olea europaea 'Swan Hill'

FRUITLESS OLIVE

Bloom Period: None of note

Mature Height x Spread: 20–30 × 30–40 ft (6–9 × 9–12 m)

Exposure: Full sun, reflected heat

Cold Tolerance: 20°F (-7°C)

Origin: Mediterranean region

Attractive gray-green leaves with silvery undersides are the hallmark of this large, evergreen tree. Fruitless olive trees are low litter as they don't produce pollen or olive fruit. This makes them ideal for the home landscape as they have all the beauty without the allergenic pollen, litter, and stains from fallen olives. In fact, in many cities, only fruitless olive trees are permitted to be planted. 'Swan Hill' is the most popular variety of fruitless olive. This slow to moderate growing tree and its wide canopy can be utilized as the primary shade tree for the front or backyard. It can also be pruned to keep it at a smaller size, making this a versatile tree for dry climate gardens. Olive trees produce many suckers, which can be removed from spring through fall. In spring, prune to your desired shape.

Parkinsonia florida

BLUE PALO VERDE

Bloom Period: Spring

Mature Height x Spread: 30 × 30 ft (9 × 9 m)

Exposure: Full sun, reflected heat

Cold Tolerance: 20°F (-6°C)

Origin: Southwest United States

One of the most iconic trees of the desert, the blue palo verde, has a green trunk and branches that can photosynthesize along with the foliage. Semi-evergreen, this desert native has small leaflets in a pleasing pale sage green color. It grows naturally in a large shrub form but is frequently trained into a tree form for landscape use. Yellow flowers cover the entire tree beginning in mid-spring and extending into early summer, creating seasonal litter. Multi-trunk specimens allow more focus on the attractive green trunk and are more resistant to wind damage than single-trunk trees. Blue palo verde has a moderate growth rate, attracts bees when in flower, and is a bird haven. Don't plant in the grass and prune in spring to maintain tree form. *P. microphylla* is a smaller palo verde tree option.

TREE PROFILES

Parkinsonia × 'Desert Museum'

'DESERT MUSEUM' PALO VERDE

Bloom Period: Spring, Summer

Mature Height x Spread: 20–30 × 35 ft (6–9 × 11 m)

Exposure: Full sun, reflected heat

Cold Tolerance: 20°F (-7°C)

Origin: Southwest United States

This tree is a hybrid of three different palo verde trees with a deeper green trunk and branches, larger yellow flowers, and is thornless. 'Desert Museum' is considered the most beautiful of palo verde trees and has a moderate to fast growth rate. The primary blooming season begins in mid-spring and goes into early summer with trees covered in golden yellow flowers. Smaller amounts of blooms appear intermittently through summer into fall. Multi-trunk forms of this tree are preferred as they are more resistant to wind damage. Avoid overwatering, which can lead to rapid growth and weak wood formation. Irrigate infrequently and deeply to promote deep root growth. Don't plant in a lawn. Prune to encourage shape tree in spring but avoid over-pruning.

Pistacia lentiscus

MASTIC

Bloom Period: Spring

Mature Height x Spread: 15–20 × 15–20 ft (5–6 × 5–6 m)

Exposure: Full sun, filtered shade

Cold Tolerance: 15°F (-9°C)

Origin: Southern Europe

The dark green foliage of the mastic tree is beautiful and adds a Mediterranean feel to the landscape. Evergreen in regions with mild winters, mastic trees have a natural shrub shape and can be used to create a tall, informal hedge or privacy screen. Another widespread use is when pruned into a tree form as their relatively compact size makes them suitable for smaller areas such as patio trees or in a side yard. Mastic trees have a slow rate of growth, but their beauty and the fact that they are low litter make them worth the wait. These trees have separate male and female types. The male trees produce small red flowers in spring, while the female has tiny red fruits in fall that turn black. Prune in spring to maintain the desired shape.

TREE PROFILES

Pistacia × 'Red Push'

'RED PUSH' PISTACHE

Bloom Period: None of note

Mature Height x Spread: 25–40 × 20–30 ft (8–12 × 6–9 m)

Exposure: Full sun

Cold Tolerance: 0°F (-18°C) (for short periods)

Origin: Hybrid of trees native to Africa, Asia, and the Middle East

The lush green foliage of this Pistache hybrid and its traditional tree shape add a mesic look to the landscape. The deciduous foliage consists of dark green leaflets that are dark red in color when they first emerge. 'Red Push' pistache trees do well when planted in a lawn or rock mulch. As temperatures cool, the leaves turn yellow to deep red. In colder regions, the autumn color occurs in fall, and in others, you'll see color in winter. The growth rate of 'Red Push' pistache starts slowly but speeds up after the first few years, so be sure to allow adequate room for it to grow. Prune in late winter to maintain its tree shape.

Prosopis glandulosa

HONEY MESQUITE

Bloom Period: Spring

Mature Height x Spread: 25 × 30–40 ft (8 × 9–12 m)

Exposure: Full sun

Cold Tolerance: 0°F (-18°C)

Origin: Mexico, Southwest United States, California

This deciduous tree is prized for its attractive branch architecture. Multi-trunk specimens showcase the twisting and curving branches to significant effect. Bright green foliage consists of narrow leaflets that add a lovely lushness to the garden. Honey mesquite is fast growing and has a wide canopy that provides filtered shade. Creamy yellow catkin flowers appear in spring, followed by long, narrow seedpods. The flowers attract bees. While this is a thorny tree, there are thornless cultivars. Like most mesquites, honey mesquite is highly drought tolerant once established and can thrive on monthly irrigation in summer. Use as the main tree in the front or backyard, planting groundcovers underneath that will appreciate the filtered shade. Prune in spring to maintain a multi-trunk shape. Avoid planting in a lawn.

TREE PROFILES

Quercus fusiformis

ESCARPMENT LIVE OAK

Bloom Period: None of note

Mature Height x Spread: 50 × 50 ft (15 × 15 cm)

Exposure: Full sun

Cold Tolerance: -10°F (23°C)

Origin: Mexico, Texas

This oak has the characteristic deep green foliage and rough bark that oaks are known for. However, this species is better adapted to hot, arid conditions than other types of oaks. Deep green leaves cover dark brown branches year-round, except for regions that experience cold winters where it is deciduous. Like most oaks, escarpment live oak has a slow growth rate but will get very large at maturity, so don't plant too close to structures. There is seasonal litter when acorns drop from the tree. This tree is suitable for bigger spaces and can be grown in a lawn. Prune to keep the traditional tree shape and to remove any root suckers. Avoid thinning the canopy of the tree.

Ulmus parvifolia

CHINESE ELM

Bloom Period: None of note

Mature Height x Spread: 40–60 × 40–60 ft (12–18 × 12–18 m)

Exposure: Full sun

Cold Tolerance: 0°F (-18°C)

Origin: China

Lush green leaves and a wide canopy make this the quintessential shade tree for dry climate regions. Branches slightly arch downward, lending to a weeping appearance. The trunk has attractive coloration ranging from pale gray to light tan. This elm species has been proven to thrive in dry climate regions but does require regular water. Chinese elm is evergreen in frost-free climates but will lose its leaves in winter in colder areas. While many elm species are susceptible to disease, Chinese elm has proven resistant, hence their popularity. Use where deep shade is desired, including in lawns. Be sure to allow room for it to grow as it will grow quite broad, so keep it away from nearby structures. Prune in late winter or spring.

Vitex agnus-castus

CHASTE TREE

Bloom Period: Summer

Mature Height x Spread: 25 × 25 ft (8 × 8 m)

Exposure: Full sun

Cold Tolerance: -10°F (-23°C)

Origin: Southern Europe

This is a large, deciduous shrub commonly grown as a multi-trunk tree. The aromatic leaves are deep green and are made of leaflets. Chaste trees have a slow to moderate growth rate and produce 3-inch (8 cm) spikes of flowers off and on through the summer that attract bees. The most common flower color is purple, but pink and white cultivars are available. Once flowers fade, small brown seed heads remain. The amount of flowering is related to heat; the hotter it is, the more they flower. Use in areas where other plants provide winter interest as the chaste tree isn't beautiful in winter. Avoid overwatering, which leads to lusher foliage but fewer flowers. Instead, focus on deep, infrequent irrigation. In spring, prune to the desired shape.

SHRUB PROFILES

Bougainvillea spp.

BOUGAINVILLEA

Bloom Period: Year-Round

Mature Height x Spread: 6–20 × 10–20 ft (2–6 × 3–6 m) (depending on the hybrid)

Exposure: Full sun, reflected heat

Cold Tolerance: 25°F (-4°C)

Origin: Brazil

Large, deep green leaves conceal thorny branches of this tropical beauty. Colorful brachts, often in shades of red or magenta, surround a tiny cream-colored flower. Bougainvillea comes in different colors and sizes, from groundcovers to shrubs and vines. While they will grow in filtered sun, flowering will be significantly decreased. In colder winter regions, flowering pauses in winter, and frost damage may occur. Bougainvillea is messy and shouldn't be planted where their debris will collect. Prune this fast-growing shrub in spring, removing any frost-damaged growth. Severe pruning can be done sporadically to remove old, woody branches. In summer, light pruning may be needed to keep it within bounds.

Caesalpinia pulcherrima

RED BIRD-OF-PARADISE, PRIDE OF BARBADOS

Bloom Period: Summer

Mature Height x Spread: 6–10 ft (2–3 m) tall and wide

Exposure: Full sun

Cold Tolerance: 10°F (-12°C)

Origin: Central America

This semi-tropical shrub has brilliant orange-red flowers that attract butterflies and hummingbirds. The foliage is medium green and made up of tiny clusters of leaflets giving it a fernlike appearance. Flowers appear continuously in summer, followed by seedpods. Be sure to allow enough room for this large shrub to reach its mature size. Suitable for use to create an informal hedge, plant against a wall, or singly as an accent plant. Excellent addition to a pollinator garden. They go dormant in the winter months and lose most of their leaves. Prune back to 1 foot (30 cm) in height in spring—they will grow back quickly.

Calliandra californica

BAJA FAIRY DUSTER

Bloom Period: Spring through Fall

Mature Height x Spread: 5 × 5 ft (1.5 × 1.5 m)

Exposure: Full sun, reflected heat

Cold Tolerance: 20°F (-6°C)

Origin: Mexico

Uniquely shaped, red flowers shaped like a feather duster are the hallmark of this Mexican native. In areas with mild winters, blooms may continue in winter. Baja fairy duster is the ultimate pollinator plant as its flowers attract butterflies and hummingbirds while other bird species enjoy the seeds within the seedpods. Deep green foliage consists of tiny leaflets clustered tightly together. It has a natural, spiky shape, making it an excellent choice for gardens with a realistic design theme. Lightly prune back in mid to late spring using hand pruners, removing up to 20 percent of its growth. Avoid severe pruning. For large shrubs, use selective thinning pruning to remove old, woody growth.

SHRUB PROFILES

Calliandra eriophylla

PINK FAIRY DUSTER

Bloom Period: Spring

Mature Height x Spread: 3 × 3 ft (1 × 1 m)

Exposure: Full sun, reflected heat

Cold Tolerance: 10°F (-12°C)

Origin: Southwestern North America

Like its cousin, Baja fairy duster, this smaller shrub has distinctive, fan-shaped flowers but in a delightful shade of pink. The primary bloom season is spring, but flowers may also come at other times of the year, including fall. Their tiny medium-green leaflets contribute to an airy, open appearance. During periods of intense cold or drought, some leaves may drop. This pink beauty thrives in desert gardens and attracts butterflies and hummingbirds. It makes a good choice for a low-growing shrub that can be used in bare areas between cacti and other succulents to provide softness. Pink fairy duster doesn't require pruning every year. In late spring, prune back by half its size every other year.

Dalea frutescens

BLACK DALEA

Bloom Period: Fall

Mature Height x Spread: 3 × 5 ft (1 × 1.5 m)

Exposure: Full sun, reflected heat

Cold Tolerance: 15°F (-9°C)

Origin: Southwestern North America

The lacy green foliage of this mounding shrub is composed of tiny, medium-green leaflets. However, don't let its delicate appearance fool you—black dalea is one tough plant. Vibrant purple flowers appear in fall, which attracts bees. In addition, they have a lovely mounded growth habit, adding an attractive softness to the landscape. Use it near spiky succulents like a desert spoon or golden barrel cacti to showcase its finely textured foliage. Black dalea also looks lovely next to the yellow-flowering turpentine bush, which also blooms in fall. Prune back every other year in spring to half its size.

SHRUB PROFILES

Dodonaea viscosa

HOP BUSH, HOPSEED BUSH

Bloom Period: Spring
Mature Height x Spread: 5–12 × 6–10 ft (1.5–4 × 2–3 m)
Exposure: Full sun, light shade
Cold Tolerance: 15°F (-9°C)
Origin: Southwest United States

This large shrub is prized for its evergreen foliage and versatility in the landscape. In spring, small, light green, papery flowers appear, which are rather unremarkable. The leaves are oblong and medium green in color. Hop bush is adaptable and serves several functions in the landscape, from a tall green backdrop along a block wall or as a large hedge. In areas too small for a tree, hop bush can be trained into a small tree. There is a purple variety, but it lacks the vigor of the actual species. Be sure to allow for its mature width. You can adjust the height of this shrub to match the function you want it to serve. Hop bush has an attractive natural shape, but it can be periodically sheared for a more traditional form in spring.

Eremophila hygrophana

BLUE BELL™ EMU

Bloom Period: Fall, Winter, Spring
Mature Height x Spread: 2–3 × 3 ft (61–91 × 91 cm)
Exposure: Full sun, reflected heat
Cold Tolerance: 17°F (-8°C)
Origin: Western Australia

The gray-blue foliage of this Australian native brings out the deeper green color of other plants in the landscape. However, its blue-violet blooms are the main attraction. Flowers appear in spring and fall, and occasionally in summer. Blue Bell™ Emu are often mistaken for Texas sage shrubs (*Leucophyllum spp.*) but bloom over a more extended period and are more compact in size, making them an excellent substitute for smaller areas. Use alongside red or yellow-flowering shrubs and groundcovers for a delightful mix of colors. They look best when grouped in threes or fives. Avoid overwatering, allowing the soil to dry out in-between watering. Little pruning is required other than light shaping in spring. Don't shear into formal shapes.

SHRUB PROFILES

Eremophila maculata

RED EMU BUSH

Bloom Period: Winter, Spring

Mature Height x Spread: 4 × 5 ft (1.2 × 1.5 m)

Exposure: Full sun, reflected heat

Cold Tolerance: 15°F (-9°C)

Origin: Australia

Beautiful, reddish-pink blooms are the hallmark of this Australian native. In mild-winter climates, flowering begins in winter, offering welcome color to the garden. Blooming continues into spring. Small, oval leaves in a pleasing deep shade of dusty green add nice background color when not in bloom. Mature shrubs have a slightly open and arching growth habit. There are several varieties of red emu, including 'Valentine', pictured above. Plant in groups of three or five for maximum impact. Prune once a year to half its size once flowering has ended in late spring. This shrub is low maintenance and beautiful but needs well-drained soil, or it will struggle.

Justicia californica

CHUPAROSA

Bloom Period: Year-round

Mature Height x Spread: 3–5 × 5 ft (91 cm–1.5 m × 1.5 m)

Exposure: Full sun, reflected heat, filtered sun

Cold Tolerance: 20°F (-6°C)

Origin: Mexico, Southwest United States

This desert native has bright green leaves that appear on gray stems. Tubular red flowers appear off and on through the year in regions with mild winters with the heaviest bloom in spring. Hummingbirds are frequent visitors to the blooms. Chuparosa can die back to the ground in cold winter locations but will grow back quickly. This shrub is best used in naturally themed landscapes where its mounded, sprawling shape is most appreciated. It is drought-deciduous and will lose its leaves when there is a lack of water, and the stems will continue to photosynthesize. Because this is a desert native, less is more when it comes to water—irrigate infrequently and deeply. Prune in spring to the desired size or cut back to the ground to rejuvenate.

SHRUB PROFILES

Justicia spicigera

MEXICAN HONEYSUCKLE

Bloom Period: Year-round

Mature Height x Spread: 3 × 5 ft (91 cm × 1.5 m)

Exposure: Filtered sun, bright shade

Cold Tolerance: 15°F (-9°C)

Origin: Mexico, Central America

Medium orange flowers decorate this small shrub throughout most of the year. The blooms are tubular in shape, which attracts hummingbirds. In areas with cold winters, flowering may pause until spring. In addition to the flowers, the lush, light green foliage adds a nice tropical feel to the landscape. Mexican honeysuckle has a naturally mounded shape. One of the desirable characteristics of this shrub is that it does well in filtered to the bright shade while handling summer heat. Don't plant in full sun, or it will struggle in low desert climates. This orange-flowering beauty is another great option for growing in a large container. Yearly pruning isn't always required. In spring, any pruning should be done by cutting it back to the ground.

Lantana camara

BUSH LANTANA

Bloom Period: Spring, Summer, Fall

Mature Height x Spread: 3–4 × 4–5 ft (90 cm–1.2 m x 1–1.5 m)

Exposure: Full sun, reflected heat

Cold Tolerance: 10°F (-12°C)

Origin: Central and South America

The dark green foliage of this mounded shrub and colorful flowers make bush lantana a popular choice for dry climate gardens. Clusters of tiny flowers form bigger blooms. The flower color is dependent on the type of hybrid and ranges from red, orange, and yellow flower clusters to pale pink and yellow. In cold winter regions, it's treated as an annual. In warmer climates, it's grown year-round. Butterflies and hummingbirds are attracted to the flowers. As blooms fade, small berry-like fruit follows, which is toxic if ingested. Use wherever you want to add a bright splash of color, including in containers. In spring, prune back to the ground to keep it compact and avoid becoming too large.

SHRUB PROFILES

Leucophyllum spp.

TEXAS SAGE, TEXAS RANGER

Bloom Period: Summer, Fall

Mature Height x Spread: 5–6 × 5–8 ft (1.5–1.8 × 1.5–2.4 m)

Exposure: Full sun, reflected heat

Cold Tolerance: 10°F (-12°C)

Origin: Texas, Mexico

Foliage in sage green and gray shades is the predominant characteristic of these large shrubs, which come in several species. Flowers are most commonly a shade of purple, but there are white and pink varieties of this shrub. The blooms come and go, but large flushes of flowers appear in response to rain or increased humidity levels. Bees are attracted to the flowers. Texas sage is semi-deciduous and loses some of its leaves in winter. Use in areas where they have room to grow, such as against a wall that receives full sun or an informal hedge. *Leucophyllum frutescens* is pictured above. However, there are many excellent species, including *L. langmaniae* and *L. zygophyllum*. Prune in spring to half their size or cut back severely to rejuvenate. Avoid shearing.

Olive europaea 'Montra'

LITTLE OLLIE® DWARF OLIVE

Bloom Period: None of note

Mature Height x Spread: 4–10 × 4–10 ft (1.2–3 × 1.2–3 m)

Exposure: Full sun

Cold Tolerance: 15°F (-9°C)

Origin: Greece, Italy

This evergreen shrub has an olive tree's desirable qualities in a shrub form. Attractive leaves with shades of gray-green on the top and pale gray underneath are the most desirable feature of this version of olive. Little Ollie® rarely flowers or produces fruit. The mature size of these shrubs is adaptable depending on the function you want them to fulfill. Keep them shorter to create a nice border or allow them to grow tall to create a wall covering or privacy screen. Pruning needs are minimal if allowed to grow tall and wide. Otherwise, prune to the desired size in spring.

SHRUB PROFILES

Ruellia peninsularis

BAJA RUELLIA, DESERT RUELLIA

Bloom Period: Year-round

Mature Height x Spread: 4 × 6 ft (1.2 × 1.8 m)

Exposure: Full sun, reflected heat

Cold Tolerance: 25°F (-4°C)

Origin: Mexico

Medium green foliage covers this nicely mounded shrub, making it a great choice as a foundation planting or filler in bare areas. The stems are pale gray. Purple flowers appear sporadically through the year, with the heaviest bloom occurring in spring. Don't confuse it with its invasive cousin (*Ruellia brittoniana*). Baja ruellia attracts butterflies and hummingbirds. Use against walls, near driveways, or sidewalks where its tolerance of reflected heat will be appreciated—plant in groups of three or more for best impact. Prune in spring to half their size (if needed) and severely prune to rejuvenate every few years.

Russelia equisetiformis

CORAL FOUNTAIN, FIRECRACKER PLANT

Bloom Period: Spring through Fall

Mature Height x Spread: 3–4 × 4–5 ft (91 cm–1.2 m × 1.2–1.5 m)

Exposure: Full sun

Cold Tolerance: 20°F (-7°C)

Origin: Mexico

Bright green foliage comprises narrow stems with a weeping shape, which adds a soft, flowing texture to the garden. The thin orange-red flowers appear throughout the warm season, attracting hummingbirds. While a coral fountain can be invasive in tropical regions, that isn't true when grown in dry climate gardens where it thrives with supplemental irrigation. Use them in high-profile areas where their beauty can be appreciated, such as near an entry, patio, or window. Coral fountain is also suitable for growing in a large container. Major pruning isn't needed every year unless there is frost damage. For mature plants, focus pruning on removing one-third of stems at the base of the plant.

SHRUB PROFILES

Senna artemisioides

FEATHERY CASSIA

Bloom Period: Mid-Winter to Spring

Mature Height x Spread: 5–6 × 5–6 ft (1.5–1.8 × 1.5–1.8 m)

Exposure: Full sun, reflected heat

Cold Tolerance: 15°F (-9°C)

Origin: Australia

The gray-green foliage of this Australian native has a finely textured, feathery appearance and shimmers on a windy day. The foliage is made up of *phyllodes*, which are very narrow leaflike structures. Fragrant, bell-shaped flowers appear in late winter into early spring, followed by tiny seedpods. Feathery cassia is an excellent option for cool-season color. The foliage makes it a great background or filler plant throughout the rest of the year—use along a fence or wall or in a staggered placement in the landscape. Prune back in mid to late spring by up to half its size once the flowers fade to remove any seedpods. Avoid pruning in summer or fall, which will reduce the flowering the following spring.

Tecoma alata (syn. Tecoma fulva subsp. Guarume)

ORANGE BELLS, ORANGE ESPERANZA

Bloom Period: Spring through Fall

Mature Height x Spread: 6–12 × 8 ft (2–4 × 2.4 m)

Exposure: Full sun

Cold Tolerance: 0°F (-18°C)

Origin: Western region of South America

Clusters of orange tubular flowers are a draw for hummingbirds and people alike. The bright green foliage adds a lush look to the landscape. One of the prized attributes of orange bells is their height, which can be used to create a tall, flowering screen in the garden. The long bloom season ensures a bountiful feast for hummingbirds while adding beauty to outdoor spaces. Orange bells have a vase-shaped growth habit with woody branches at the base transitioning to lush green foliage upward. This fast-growing shrub can be utilized to create shade for smaller areas. Prune in spring. Severe pruning can be done every few years to eliminate old, woody growth or selectively thin branches at the base to preserve its overall height.

SHRUB PROFILES

Tecoma stans var. angustata

YELLOW BELLS

Bloom Period: Spring through Fall

Mature Height x Spread: 8–10 × 8 ft (2.4–3 × 2.4 m)

Exposure: Full sun, reflected heat

Cold Tolerance: 0°F (-18°C)

Origin: Southwestern North America

The lush foliage of yellow bells makes this a popular choice for deep greens in the landscape. Large, tubular yellow flowers appear throughout the warm season, beckoning hummingbirds. Narrow seedpods follow the flowers. A closely related variety, *Tecoma stans var. stans*, has broader leaves and is more cold-tender, and hardy to 20°F (-7°C). Popular hybrids are available in orange, red, and yellow shades. This large shrub is fast-growing, making it a good choice for covering a bare wall or fence. If caterpillars are a problem, spray with an organic pesticide containing BT (*Bacillus thuringiensis*). It is frost-tender but grows back quickly in spring. Therefore, prune in spring once the danger of frost has passed. Pruning can be done by selective thinning old, woody branches or cutting back severely to 1 ft (30 cm) tall—regrowth is rapid.

Teucrium fruiticans

SHRUBBY GERMANDER

Bloom Period: Spring through Fall

Mature Height x Spread: 3–4 × 5 ft (91 cm–1.2 m × 1.5 m)

Exposure: Full sun

Cold Tolerance: 10°F (-12°C)

Origin: Mediterranean region

The foliage of this medium shrub adds welcome color contrast with darker green plants. Leaves are silvery blue on the top and white on the bottom. In spring, blue flowers appear in a big flush with sporadic blooms until fall. In hot summer regions, blooming may pause. The small flowers attract bees. Shrubby germander has a lovely natural, spiky shape that adds a nice texture. Blue-flowering plants aren't common, so this is an excellent choice for a colorful garden. Pruning needs are minimal as they don't require yearly pruning. Larger shrubs can be cut back severely in spring if needed, or light pruning can also be done at the same time.

GROUNDCOVER AND PERENNIAL PROFILES

Berlandiera lyrata

CHOCOLATE FLOWER

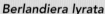

Bloom Period: Spring through Fall

Mature Height x Spread: 1 × 2 ft (30 × 61 cm)

Exposure: Full sun, filtered sun

Cold Tolerance: -30°F (-34°C)

Origin: Southwest United States

The pretty daisy-like flowers of this perennial have a surprising fragrance. As you stop to sniff the blooms, you will smell the unmistakable scent of chocolate. The flowers are yellow with maroon centers. The foliage is finely textured with a gray-green color with a clumping growth habit. Chocolate flower is a relatively informal plant and works well where a native or natural garden design is used. Place them where you can smell the fragrance, which is strongest in the morning. Periodically prune back the blooms during the growing season to stimulate new ones. In spring, severely cut back straggly specimens.

Carissa macrocarpa

NATAL PLUM

Bloom Period: Spring

Mature Height x Spread: 8 in–3 ft × 2–5 ft (20–91 cm × 61 cm–1.5 m)

Exposure: Full sun, bright shade

Cold Tolerance: 20°F (-7°C)

Origin: South Africa

The popularity of this shrub/groundcover is its dark, lush foliage, which adds a nice oasis-like feel to gardens in arid regions. In spring, white star-shaped flowers appear that are fragrant, but flowering can be reduced when used in reflected heat locations. Edible, small red fruits follow the flowers. Natal plum comes in several varieties, from small shrub forms to groundcovers. Thorns are present along its stems that are well-hidden by the foliage. This South African native is very versatile as it can be grown in either sun or shade. Use as a foundation plant underneath a window or plant in groups of three or five to create a welcome splash of deep green in the landscape. While it is slow-growing, it seldom needs pruning except to remove frost-damaged growth in spring.

GROUNDCOVER AND PERENNIAL PROFILES

Chrysactinia mexicana

DAMIANITA

Bloom Period: Spring, Fall

Mature Height x Spread: 1–2 × 2 ft (30–61 cm × 61 cm)

Exposure: Full sun

Cold Tolerance: 0°F (-18°C)

Origin: Southwest United States

This shrubby groundcover takes center stage during its two bloom seasons in the arid garden. The aromatic foliage of damianita consists of deep green, narrow leaves tightly clustered together. While its appearance is rather unremarkable when not in bloom, damianita is transformed into golden yellow mounds that resemble pillows throughout the landscape. This plant works best in naturally themed landscapes when planted in groups of three or more. Use alongside large rocks, along pathways, or for lower interest near tall shrubs—plant in well-drained soil. In fall, lightly shear back once flowering ends.

Convolvulus cneorum

BUSH MORNING GLORY

Bloom Period: Spring

Mature Height x Spread: 1–2 × 3 ft (30–61 × 91 cm)

Exposure: Full sun, reflected heat

Cold Tolerance: 15°F (-9°C)

Origin: Southern Europe

The silvery-gray foliage of bush morning glory makes it a great choice to create foliar color contrast near plants in shades of darker greens. In spring, white, bell-shaped flowers appear that add another decorative element to the landscape. Bush morning glory has a lovely mounded shape that is best highlighted alongside large rocks, steps, or overhanging from raised beds. However, this is a groundcover that can handle stressful spots that receive hot, reflected sun. While its needs are few, it must have well-drained soil to thrive—avoid overwatering. Little to no pruning is required. If needed, lightly prune back in mid-spring once the flowers have faded.

GROUNDCOVER AND PERENNIAL PROFILES

Eremophila glabra 'Mingenew Gold'

'OUTBACK SUNRISE' EMU

Bloom Period: Spring

Mature Height x Spread: 1 × 6 ft (30 cm × 1.8 m)

Exposure: Full sun, reflected heat, filtered shade

Cold Tolerance: 15°F (-9°C)

Origin: Australia

The bright green color of this low-growing plant comes from narrow leaves clustered along green stems. The green shade of this groundcover matches that of a lush green lawn. Pale yellow flowers appear in spring. 'Outback Sunrise' emu thrives in various exposures, making it highly versatile in the landscape. When planted in groups, it creates an island of green and is a visual substitute for thirsty lawns, although it cannot be walked on. While this is a low-maintenance choice for the garden, it must have well-drained soil, and be sure to avoid overwatering. Pruning needs are minimal and consist of light shearing (if needed) in spring once the flowers have faded.

Lantana montevidensis

TRAILING LANTANA

Bloom Period: Spring through Fall

Mature Height x Spread: 1–2 × 4 ft (30–61 cm × 1.2 m)

Exposure: Full sun, filtered shade

Cold Tolerance: 20°F (-6°C)

Origin: South America

Trailing lantana is a fast-growing groundcover with deep green leaves that produce blooms over a long season. Tiny purple flowers are borne in clusters that add a pretty shade of excellent color to the garden. Butterflies are attracted to their blooms. Plant in mass for a splash of green to create a visually cooling effect or use in containers where its trailing growth habit can be enjoyed. Trailing lantana looks excellent when planted alongside large rocks and spiky succulents such as agave. A white flowering form is available, which can be mixed in with the more common purple type to create a pretty contrast of colors. Lantana is resistant to deer and rabbits. Prune severely every other spring to rejuvenate.

GROUNDCOVER AND PERENNIAL PROFILES

Lantana x 'New Gold'

GOLD LANTANA

Bloom Period: Spring through Fall

Mature Height x Spread: 2–3 × 3–4 ft (61–91 cm × 91 cm–1.2 m)

Exposure: Full sun, reflected heat

Cold Tolerance: 10°F (-12°C)

Origin: South America

Golden yellow flowers decorate the lush green foliage of this shrubby groundcover. 'New Gold' lantana thrives in the intense heat of dry climate gardens, where it's one of the most popular groundcovers. The soft mounded shape is ideal for creating softness in the garden, whether used alongside large rocks or to soften the edges of garden beds. Plant in groups of three or more to maximize the color impact of the blooms. Like all lantana species, 'New Gold' makes an excellent container plant and can be treated as a flowering annual in cold regions during the summer. Use for lower interest in front of tall shrubs or high-profile spots throughout the landscape. 'New Gold' lantana attracts butterflies and is resistant to deer and rabbits. Prune severely every other spring to rejuvenate. In other years, prune back by half its size.

Lavandula stoechas

SPANISH LAVENDER

Bloom Period: Spring

Mature Height x Spread: 2 × 3 ft (61 × 91 cm)

Exposure: Full sun

Cold Tolerance: 10°F (-12°C)

Origin: Spain

The beauty and fragrance of lavender are a welcome addition to dry climate gardens. However, some species of lavender can struggle. Spanish lavender is native to Spain's dry, arid region, which makes it the best choice for areas with hot, dry summers. It has a more compact growth habit than other lavenders with narrow, gray-green leaves. In spring, violet blooms appear that add fragrant beauty to the garden. Plant in groups of three by a front entry where its fragrance will be enjoyed year-round, or plant in a pot. Lavender needs well-drained soil to grow its best, and avoid planting in locations with reflected heat. Remove spent flowers and lightly prune in spring, only if required.

GROUNDCOVER AND PERENNIAL PROFILES

Melampodium leucanthum

BLACKFOOT DAISY

Bloom Period: Spring through Fall
Mature Height x Spread: 1 × 2 ft (30 × 61 cm)
Exposure: Full sun, part shade
Cold Tolerance: -20°F (-29°C)
Origin: Southwest United States, Mexico

This perennial's pretty, white daisy flowers add a cottage-like feel to dry climate gardens. Dark green, narrow leaves create a spreading, mounded shape. In hot desert locations, blooms appear in spring and fall, while in milder climates, flowering can also occur in summer. Blackfoot daisy may look delicate, but it easily handles the extremes of dry climate gardens, from temps over 110°F (43°C) to frigid winters. They appear as dead sticks in cold winters but rapidly leaf out when the weather warms. Plant in groups of two or more where its blooms can be appreciated up close. It makes a great pairing next to boulders and succulents. Take care to plant in well-drained soil and don't overwater. Pruning needs are minimal and consist of removing old woody growth in spring.

Penstemon eatonii

FIRECRACKER PENSTEMON

Bloom Period: Winter, Spring, Summer
Mature Height x Spread: 1 × 2 ft (30 × 61 cm)
Exposure: Full sun
Cold Tolerance: -20°F (-29°C)
Origin: Southwest United States

Penstemons are the flowering beauties of the western half of North America. They come in many different species, with several that excel in regions with hot, dry summers. Firecracker penstemon has a clumping growth habit made up of medium-green leaves. In mid-winter, spikes of orange-red flowers appear 2 to 3 feet (61–91 cm) above the leaves, adding vibrant color to the cool season landscape. In cold winter regions, blooming will occur in summer. Hummingbirds are attracted to the blooms. Use them in high-profile areas where their beauty can be viewed up close. Firecracker penstemon looks great near boulders, and they do self-seed. Well-drained soil is crucial to their success, and avoid overwatering. Prune away flowering spikes once they begin to dry.

GROUNDCOVER AND PERENNIAL PROFILES

Penstemon parryi

PARRY'S PENSTEMON

Bloom Period: Spring

Mature Height x Spread: 1 × 1 ft (30 × 30 cm)

Exposure: Full sun

Cold Tolerance: 15°F (-9°C)

Origin: Arizona, Mexico

Pink flowering spikes decorate gardens wherever this penstemon is planted. Despite their delicate appearance, Parry's penstemon thrives in hot sunny regions. The foliage is made up of gray-green leaves that are rather unremarkable. In spring, 2–3 foot (61–91 cm) tall stems are covered in pretty pink flowers that attract hummingbirds. This is an excellent choice for those who desire a cottage garden feel or love the look of wildflowers. Parry's penstemon does reseed. Use it in places where you can view their stunning flowers and the hummingbirds who visit them. Be sure to plant in well-drained soil. Penstemon can struggle if you give them too much water. Prune back flowering spikes to the base of the plant once they begin to dry.

Rosmarinus officinalis 'Prostratus'

TRAILING ROSEMARY

Bloom Period: Winter, Spring

Mature Height x Spread: 2 × 4 ft (61 cm × 1.2 m)

Exposure: Full sun, part shade

Cold Tolerance: 10°F (-12°C)

Origin: Mediterranean region

This trailing form of rosemary has all the qualities of shrublike forms. The intensely aromatic foliage is made up of thin leaves that are deep green on the top and gray-green on the bottom. In winter and spring, small light blue flowers appear that attract bees. Rosemary is available in shrub and groundcover varieties. Use in areas where its attractive foliage will be appreciated. Prostrate forms of rosemary are an excellent choice for trailing over a wall. The foliage is edible and can be used to flavor your favorite dishes. Pruning of rosemary consists of removing dead wood from underneath the plant in spring. Don't prune it back severely as it may not grow back.

GROUNDCOVER AND PERENNIAL PROFILES

Salvia greggii

AUTUMN SAGE

Bloom Period: Spring through Fall

Mature Height x Spread: 3 × 3 ft (91 × 91 cm)

Exposure: Full sun (part shade, low desert)

Cold Tolerance: 0°F (-18°C)

Origin: Mexico, Texas

Vibrantly colored blooms of this shrubby perennial are a favorite of hummingbirds. The foliage is aromatic with medium-green leaves borne along woody stems. Autumn sage comes in several flower colors, but red and pink are the most common. In regions with scorching summers, flowering may pause until the weather cools again in the fall. Plant underneath windows to view the flowers and hummingbirds up close, or plant near a patio or courtyard area. For low desert regions, plant them in locations where they will receive filtered shade. Lightly shear back in spring before blooming begins.

Sphaeralcea ambigua

GLOBE MALLOW

Bloom Period: Spring, Fall

Mature Height x Spread: 3 × 4 ft (91 cm × 1.2 m)

Exposure: Full sun, reflected heat

Cold Tolerance: -10°F (-23°C)

Origin: Southwest United States

Colorful blooms are borne above the gray-green foliage of this shrubby perennial. The flowers are cup-shaped and are a favorite of bees. The most common bloom color is orange. However, other flower colors are available, including white, pink, lavender, and red. Globe mallow has a natural round growth habit, and flowering spikes cover the foliage when in bloom, making it an excellent choice for those who desire a cottage-garden look. Use caution when pruning as the foliage is irritating to the skin. Globe mallow is easiest to grow from seed, and it will self-seed in the landscape. Prune back by half its size in late spring once flowering has ended.

Tetraneuris acaulis
(syn. Hymenoxys acaulis)

ANGELITA DAISY

Bloom Period: Spring through Fall (year-round in mild Winter areas)

Mature Height x Spread: 1 × 2 ft (30 × 61 cm)

Exposure: Full sun, reflected heat

Cold Tolerance: -20°F (-29°C)

Origin: Western North America

Despite its petite size, this perennial makes a large impact in the garden. Golden yellow daisy-like flowers fan out over the plant on 6-inch (15 cm) tall stems over dark green, clumping foliage. Although it looks delicate, angelita daisy easily handles intense heat and severe cold. Flushes of yellow flowers appear off and on throughout most of the year in mild winter regions. Angelita daisy works best planted in groups of two or three, which help magnify its impact. Scatter several groupings throughout the garden to tie it together visually. Periodically remove spent blooms throughout the year. In spring, prune back to the ground every three years to get rid of old, woody growth and rejuvenate.

VINE PROFILES

Antigonon leptopus

QUEEN'S WREATH, CORAL VINE

Bloom Period: Summer through Fall

Mature Height x Spread: 15–25 × 15–25 ft (4.6–7.6 × 4.6–7.6 m)

Exposure: Full sun

Cold Tolerance: 20°F (-7°C)

Origin: Mexico

Delicate sprays of pink flowers decorate this Mexican native through hot summers. The foliage of queen's wreath is lush, light green with an attractive heart shape. While invasive in tropical regions, the queen's wreath is well-behaved in dry climates. Plant at the base of vertical surfaces such as walls or fences, providing a trellis for support. Bees and butterflies are frequent visitors to queen's wreath. This fast-growing vine is completely deciduous, meaning it will lose all its leaves in winter. Prune back to the ground in spring once the threat of frost is over, and it will grow back quickly.

Bignonia capreolata

CROSSVINE

Bloom Period: Spring

Mature Height x Spread: 30 × 30 ft (9 × 9 m)

Exposure: Full sun, part shade

Cold Tolerance: -10°F (-23°C)

Origin: Southeastern United States

The lush green foliage of this vine adds a lovely, vertical backdrop to dry climate gardens. Red, tubular flowers appear in spring that attract hummingbirds. 'Tangerine Beauty' is a popular variety with orange/pink blooms. Crossvine is an excellent alternative to trumpet vine (*Campsis radicans*) as it isn't as invasive and easier to control. Use to cover a bare wall, fence, or chain-link fence. Be sure to allow enough room for it to grow and keep other plants out of reach of its tendrils. Provide support such as a trellis to help train it up and outward. Like most vines, crossvine needs a couple of years before it begins to grow. Prune as necessary to control its growth and fertilize in spring. Light pruning can be done in summer.

Gelsemium sempervirens

CAROLINA JASMINE

Bloom Period: Winter, Spring

Mature Height x Spread: 15–20 × 15–20 ft (4.6–6 × 4.6–6 m)

Exposure: Full sun, part shade

Cold Tolerance: 15°F (-9°C)

Origin: Southeastern United States

Attractive green foliage covers the maroon stems of this vine throughout the year. In spring, fragrant yellow flowers appear that add warm shades of color to the garden. Carolina jasmine is the state flower of South Carolina and is a great choice for landscapes where deer and rabbits are a problem as it is resistant to them. The entire vine is toxic if ingested, and its sap can cause skin irritation—plant on walls or fences where beauty and privacy are desired. Avoid planting on west-facing walls or where reflected heat is present. Provide a trellis for support. Fertilize in spring and prune once flowers drop after blooming.

VINE PROFILES

Hardenbergia violaceae

PURPLE LILAC VINE

Bloom Period: Winter

Mature Height x Spread: 15 × 20 ft (4.6 × 6 m)

Exposure: Full sun, part shade

Cold Tolerance: 20°F (-7°C)

Origin: Australia

This vine's dark green, narrow leaves are beautiful and add a lovely backdrop wherever it is used. From late winter to early spring, sprays of purple flowers appear, which add a very welcome splash of color when not much else is blooming. The blooms resemble those of lilac but aren't fragrant. While the bloom period only lasts four to five weeks, it's well worth it. The foliage adds an attractive vertical element throughout the rest of the year. Use to cover a bare wall or pool fence. Purple lilac vine is an excellent choice on walls across from a window where its beauty can be appreciated. Provide a trellis for support. Prune to the desired size in spring.

Macfadyena unguis-cati

CAT CLAW VINE

Bloom Period: Spring

Mature Height x Spread: 25 × 25 ft (7.6 × 7.6 m)

Exposure: Full sun

Cold Tolerance: 20°F (-7°C)

Origin: Central and South America

This heat-loving vine is excellent for growing along vertical walls or fences that experience the full sun and reflected heat. It's one of the few that can tolerate west-facing exposures against a wall. The medium-green foliage of this vine is arranged with tendrils that end in a tendril that resembles a cat's claw. Beautiful, yellow flowers appear in spring, followed by a long skinny seedpod. While cat claw vine is an excellent choice for full sun areas, care must be taken as it is very vigorous and can quickly overtake nearby plants and structures. No trellis is required. In spring, prune away frost-damaged growth and then focus pruning efforts to keep it restrained. You can prune it back to the ground every few years to check its spread.

Pandorea jasminoides

PINK BOWER VINE

Bloom Period: Spring, Fall

Mature Height x Spread: 15–20 × 15–20 ft (4.6–6.1 m × 4.6–6.1 m)

Exposure: Full sun, part shade

Cold Tolerance: 25°F (-4°C)

Origin: Australia

The true beauty of this vine lies in its tubular flowers. Pale pink blooms with deep red centers are clustered over lush green foliage. While most flowering occurs in spring, some repeat blooming can occur in summer and fall. Pink bower vine is slow to moderate growth and is frost tender; however, it recovers quickly in spring. In low desert gardens, this pink beauty is best used in areas with part shade or that receive morning sun only. It will require a trellis or other support. Prune in early spring to remove any frost-damaged growth and to cut back any wayward tendrils. Fertilize in spring.

VINE PROFILES

Podranea ricasoliana

PINK TRUMPET VINE

Bloom Period: Spring, Summer, Fall

Mature Height x Spread: 20 × 10 ft (6 × 3 m)

Exposure: Full sun, part shade

Cold Tolerance: 25°F (-4°C)

Origin: South Africa

Large sprays of pink flowers cover the deep green foliage of this shrubby vine over a long season. In hot summer regions, flowering may pause in summer and resume in fall. This vine has a lush appearance yet does best with infrequent, deep watering. While invasive in areas with abundant rainfall, it doesn't have this problem in dry climate regions where its growth is controlled. Pink trumpet vine's growth habit is a combination of shrub and vine. When first planted, use a trellis or stake to help train it upward. Avoid planting on west-facing walls in low desert gardens. Prune to desired shape and size in spring once the threat of freezing temperatures has passed. It can be severely pruned if needed every few years.

Rosa banksiae

LADY BANKS' ROSE

Bloom Period: Winter, Spring

Mature Height x Spread: 20 × 15 ft (6 × 4.6 m)

Exposure: Full sun, part shade

Cold Tolerance: -10°F (-23°C)

Origin: China

This rose climber is a great low-maintenance choice for dry climate gardeners. The glossy green leaves are narrower than traditional roses, and the stems are thornless. Lady Banks' blooms once a year in spring when its branches are covered with fragrant white or yellow roses, depending on the variety. This versatile rose can be trained upright on a trellis as a vine or grown as a large shrub with arching stems. While it can take a few years for Lady Banks' to fill out, it is well worth the wait. Prune yearly, once the blooming has ended, in mid to late spring. Focus pruning efforts to promote upright and outward growth while cutting back any wayward branches. Don't prune later in the year, or there will be far fewer roses the following year.

Trachelospermum jasminoides

STAR JASMINE

Bloom Period: Winter, Spring

Mature Height x Spread: 15–25 × 15–25 ft (4.6–7.6 × 4.6–7.6 m)

Exposure: Full sun (part shade, low desert)

Cold Tolerance: 5°F (-15°C)

Origin: Japan

The foliage of this jasmine vine is very dark green in color with rounded leaves that grow on dark brown stems. In spring, white, star-shaped blooms appear with jasmine's intoxicating, sweet scent. It can be grown as a vine or a groundcover. In regions with mild summer, star jasmine can grow in full sun, but avoid areas with reflected heat; for gardens with hot summers, plant in the filtered sun for best results. Star jasmine can grow in the shade to provide upright interest but won't flower much, if at all. Use where its fragrance can be appreciated. Provide a trellis for support if grown as a vine. Fertilize in spring and prune once the flowers drop. If used as a groundcover, do major pruning in spring and light cutting back in summer, if needed.

CACTI/SUCCULENT PROFILES

Agave x 'Blue Glow'

'BLUE GLOW' AGAVE

Bloom Period: None of note

Mature Height x Spread: 2 × 3 ft (61 × 91 cm)

Exposure: Full sun, part shade

Cold Tolerance: 15°F (-9°C)

Origin: Hybrid of agaves native to Mexico

This agave is a hybrid with two parents, *A. attenuata* and *A. ocahui*. The main feature of this hybrid is the coloration along the edges of its blue-gray leaves, which consists of a narrow band of red and then yellow. When the light hits the agave at the right angle, the edges appear to glow. Plant in groups, arranged in straight lines, or add in your favorite container. Its compact size makes it a good choice for smaller spaces in the garden, such as a courtyard or patio area. In low desert climates, avoid planting in places where it will receive full afternoon sun, which may burn the leaves. 'Blue Glow' agave doesn't produce offsets and lives approximately ten years or more, before flowering at the end of its life—water every two weeks in the summer months.

Agave geminiflora

TWIN FLOWER AGAVE

Bloom Period: None of note

Mature Height x Spread: 2 × 3 ft (61 × 91 cm)

Exposure: Full sun, reflected heat, part shade

Cold Tolerance: 15°F (-9°C)

Origin: Mexico

The thin succulent leaves of this agave give it a finely textured appearance compared to most other agave species. Curly white filaments are attached to the edges of the dark green leaves, adding another decorative element to the twin flower agave. Besides its attractiveness, twin flower agave can grow in various exposures, including shade, where many succulents struggle. It is a good choice for contemporary landscape design when planted in straight rows or staggered, highlighting its narrow leaf structure. Like most small- to medium-sized agave, twin flower agave makes an excellent container plant. It doesn't produce offsets and lives approximately ten years before flowering at the end of its life.

CACTI/SUCCULENT PROFILES

Agave parryi v. truncata

ARTICHOKE AGAVE

Bloom Period: None of note

Mature Height x Spread: 2–3 × 3 ft (61–91 × 91 cm)

Exposure: Full sun, part shade

Cold Tolerance: 10°F (-12°C)

Origin: Mexico

This agave has a beautiful rosette arrangement of leaves that arch inward to resemble an artichoke. Each leaf is a lovely blue-gray and edged with maroon "teeth" along the sides, terminating in a single spike at the tip. The shape and color make it a great succulent to pair alongside groundcovers with deep green foliage to highlight the distinct textures. Alternatively, use it in straight rows to create an attractive geometric look, or plant it in your favorite container. Artichoke agave does well in full sun and part shade, but its leaves will spread out more in the shade. After an average of ten to twenty years, it will produce a flowering stalk at the end of its life. Provide supplemental water every three to four weeks, spring through early fall. Remove offsets as they are produced by cutting the fleshy root that connects them to the parent plant and give away or replant elsewhere.

Agave vilmoriniana

OCTOPUS AGAVE

Bloom Period: None of note

Mature Height x Spread: 4 × 5 ft (1.2 × 1.5 m)

Exposure: Full sun (part shade in the low desert)

Cold Tolerance: 20°F (7°C)

Origin: Mexico

The medium-green leaves of this Mexican native give a Medusa-like shape to the landscape. Octopus agave doesn't produce offsets, and the edges of its leaves are smooth with a single spine at the end. The leaves bring welcome curves to the succulent genre of plants. This agave is relatively short-lived and usually flowers within ten years. The flowering stalk has hundreds of baby agave ready to be planted in the garden or given to friends. Provide supplemental water every three weeks, spring through early fall. It does well in full sun and part shade, increasing its versatility in the garden; however, filtered shade is best in low desert regions.

CACTI/SUCCULENT PROFILES

Agave weberi

WEBER AGAVE

Bloom Period: None of note

Mature Height x Spread: 5 × 6–8 ft (1.5 × 1.8–2.4 m)

Exposure: Full sun, part shade

Cold Tolerance: 10°F (-12°C)

Origin: Southwest United States, Mexico

The relatively big size of weber agave makes it a good option for areas where a larger succulent is needed for height and especially the width. The gray-green leaves of this agave have a smooth appearance but have small "teeth" along the edges and end in a very sharp point. It is best utilized in areas with plenty of room to grow to its mature size, such as near large rocks or on the edge of a gently contoured mound. Weber agave can also be a focal point, mainly when surrounded by flowering groundcovers or lower growing succulents. A shorter-lived agave, it will flower at approximately six to eight years at the end of its life. It does produce some offsets that should be removed as they appear. The sharp tips of the leaves can be clipped off if desired.

Aloe vera syn. Aloe barbadensis

MEDICINAL ALOE

Bloom Period: Winter, Spring

Mature Height x Spread: 2 × 3–5 ft (61 × 91 cm–1.5 m)

Exposure: Full sun, part shade

Cold Tolerance: 25°F (-4°C)

Origin: Africa

Prized for the soothing qualities of its gel-like sap, aloe vera also adds beauty to the garden. Its spiky gray-green leaves have small teeth along the margins. In winter, 3-foot (61 cm) spikes begin to appear, topped with columnar, yellow blooms that hummingbirds enjoy feeding on. As this aloe matures, it produces other clumps of leaves that increase its width. Use aloe vera underneath trees for a spiky element or against a wall where its shape can be appreciated. To keep aloe looking neat, dig up the aloe and divide; discard excess clumps and replant a healthy section in the original spot. Aloe grown in the desert appreciates afternoon shade or part shade.

CACTI/SUCCULENT PROFILES

Aloe × 'Blue Elf'

'BLUE ELF' ALOE

Bloom Period: Winter, Spring

Mature Height x Spread: 18 in × 2 ft (46 × 61 cm)

Exposure: Full sun, part shade

Cold Tolerance: 15°F (-9°C)

Origin: South Africa

This little aloe, with its pretty blooms, thrives in arid regions. Narrow, gray-green leaves have small "teeth" along the edges and grow in clumps. Tubular orange flowers appear in winter to early spring, much to the delight of hummingbirds. 'Blue Elf' gradually expands in width as it ages. It does well in sun or part shade, allowing its use in many locations. Plant in groups of three or more straight rows or stagger the aloes in a more natural pattern. Single specimens look great when planted near large rocks or in containers. Dig up the entire aloe every three years and break off a clump with roots and replant in the same spot to keep them neat. Discard the rest or give it to friends. Every year, prune off dead flower stalks at the base of the plant.

Aloe ferox

CAPE ALOE

Bloom Period: Winter, Spring

Mature Height x Spread: 5 × 3 ft (1.5 m × 91 cm)

Exposure: Full sun, part shade

Cold Tolerance: 25°F (-4°C)

Origin: South Africa

This statuesque aloe has large succulent leaves that range from 1 to 3 feet (30–91 cm) in length, which are covered in tiny spines. The vibrant orange/red blooms are borne on stalks that stand 2 to 4 feet (61 cm–1.2 m) above the leaves from winter into spring. Cape aloe is an excellent choice for an accent plant that is a central focal point due to the size of its leaves and unique flowers. Hummingbirds flock to drink and appreciate nectar from the blooms. Plant where its unique beauty can best be seen out of a window, courtyard, or patio setting. It can also be grown in a large container. In low desert regions, provide filtered sunlight or afternoon shade to prevent the burning of the leaves. Prune back the flowering stalk once the blooms have faded.

CACTI/SUCCULENT PROFILES

Dasylirion wheeleri

DESERT SPOON, SOTOL

Bloom Period: Summer

Mature Height x Spread: 5 × 5–6 ft (1.5 × 1.5–1.8 m)

Exposure: Full sun, reflected heat

Cold Tolerance: 0°F (-18°C)

Origin: Southwest United States, Mexico

This hardy succulent is known for its fan-shaped growth habit that adds a fine-textured, spiky look to the landscape. Narrow, gray-green leaves are arranged in a rosette pattern. Each leaf has small teeth along the edges, so locating desert spoon away from high traffic areas is best. The tips of the leaves turn tan, which is normal. As it grows, it forms a short trunk. Mature specimens can produce a flowering stalk each summer that can be cut back in fall. Plant desert spoon alongside columnar cacti where their distinct shapes will contrast nicely together or inhospitable landscape areas where hot, reflected heat is present. Prune back lower leaves as they turn brown. Don't over-prune into a V-shape, which isn't healthy for the plant.

Echinocactus grusonii

GOLDEN BARREL

Bloom Period: Spring, Summer

Mature Height x Spread: 2–3 × 2.5 ft (61–91 × 76 cm)

Exposure: Full sun, part shade

Cold Tolerance: 20°F (-7°C)

Origin: Mexico

This barrel cactus's rounded shape and golden yellow spines make it a massive favorite in dry climate gardens. This cactus has a nice medium green color with the typical vertical "ribs" that run up and down the sides. The arching spines are borne in clusters along the ribs and cover the cactus in a lovely shade of yellow. Like many other cacti, golden barrels do well in full sun. However, they can also tolerate part shade. Plant in groups of three or more for an eye-catching groundcover near boulders or underneath the filtered shade of a tree. Additionally, golden barrels make excellent container plants by themselves or paired with other succulents. Provide supplemental irrigation in low desert gardens spring through fall.

CACTI/SUCCULENT PROFILES

Euphorbia antisyphilitica

CANDELILLA

Bloom Period: Spring

Mature Height x Spread: 1–2 × 2–3 ft (30–61 × 61–91 cm)

Exposure: Full sun, reflected heat

Cold Tolerance: 10°F (-12°C)

Origin: Mexico, Southwest United States

Here is a succulent that thrives in harsh places where the full sun and radiated heat are the norms. Candelilla is an upright succulent with gray-green stems that grow in a large clump. The gray tint of the stems comes from its waxy coating. It gradually expands in size with new stem growth from the base as it matures. They are extremely drought tolerant, needing monthly watering in summer in desert gardens. Tiny, pale-pink flowers may appear along the stems in spring but disappear after a few days. They have a milky sap that is irritating to the skin and eyes. Prune excess stems from the base as needed to control the width.

Euphorbia lomelii (syn. Pedilanthus macrocarpus)

LADY'S SLIPPER, SLIPPER PLANT

Bloom Period: Spring, Fall

Mature Height x Spread: 3 × 3 ft (91 × 91 cm)

Exposure: Full sun, part shade

Cold Tolerance: 25°F (-4°C)

Origin: Mexico

This succulent has smooth medium-green stems that are usually leafless. Lady's slipper have a nice upright growth habit. Small red-orange flowers are produced at the tip of each stem that attracts hummingbirds. When grown in full sun, the stems are generally upright but curve with a unique Medusa-like shape in the shade. While lady's slipper does well in full sun and filtered shade, avoid using it in areas with hot, reflected sun. They are excellent for courtyard and patio plantings and do equally well in containers. Lady's slipper is also a great option for planting next to large rocks. The milky sap of this succulent is irritating to the skin and eyes. Prune excess stems from the base as needed to control the width.

CACTI/SUCCULENT PROFILES

Euphorbia resinifera

MOROCCAN MOUND

Bloom Period: Spring

Mature Height x Spread: 1 × 3–5 ft (30 cm × 90 cm–1.5 m)

Exposure: Full sun, part shade

Cold Tolerance: 20°F (-7°C)

Origin: Africa

This succulent is often mistaken for a cactus due to its upright stems and clumping growth habit. Each stem is light green and has four distinct ridges with tiny thorns. As the Moroccan mound matures, it gradually grows outward, producing additional stems. Tiny yellow flowers may appear in spring, particularly in milder climates. Because it spreads as it grows, it can be used as a groundcover in the garden. Plant a single specimen next to boulders for a decorative combination of textures, or use several in straight rows or a staggered arrangement. Moroccan mound has a milky sap that irritates the eyes and skin. Provide supplemental water using a succulent watering schedule for your area. Little to no pruning is required.

Euphorbia rigida (syn. E. biglandulosa)

GOPHER PLANT, SILVER SPURGE

Bloom Period: Late Winter, Spring, Summer

Mature Height x Spread: 2–3 × 3 ft (61–91 × 91 cm)

Exposure: Full sun

Cold Tolerance: -20°F (-29°C)

Origin: Mediterranean region

Despite their name, gopher plants are attractive succulent groundcovers with narrow, pointed gray-green leaves arranged along arching stems. In late winter into spring, and summer in cold winter regions, chartreuse-colored flowers appear. As the flowers age, they turn color and are very pretty. Gopher plants look best in full sun and *must have* well-drained soil. Use as a groundcover by planting several throughout the landscape near large rocks or use in a mass planting. It has a milky sap that is irritating to the skin and eyes. Provide supplemental irrigation using succulent watering guidelines for your area. Don't overwater. Prune away spent flowering stems to the base of the plant once flowers fade.

CACTI/SUCCULENT PROFILES

Euphorbia tirucallii 'Sticks on Fire'

FIRESTICKS, RED PENCIL TREE

Bloom Period: None of note

Mature Height x Spread: 4–8 × 3–5 ft (1.2–2.4 × 91–1.5 m)

Exposure: Full sun

Cold Tolerance: 30°F (-1°C)

Origin: Africa

This colorful succulent has vibrant orange-red colored stems that are green at the base. In addition to its color, its thin stems and spiky growth habit ensure that it stands out wherever its planted. The coloring is caused by a lack of chlorophyll in the stems. While 'Firesticks' start small, they can grow up to 6 to 8 feet (1.8–2.4 m) tall, so be sure to allow plenty of room for them to grow. Avoid planting in the shade, which will decrease the orange/red color. Use single specimens as a focal point in the landscape or an informal hedging plant. This succulent is frost-tender, and protection is needed when temps hover around 32°F (0°C). Its milky sap is irritating to the eyes and skin. Use it in the front yard and away from foot traffic. Water once to twice a month, spring through early fall.

Fouquieria splendens

OCOTILLO

Bloom Period: Spring

Mature Height x Spread: 15–20 × 8–10 ft (4.6–6.1 × 2.4–3 m)

Exposure: Full sun, reflected heat

Cold Tolerance: 0°F (-18°C)

Origin: Southwest United States, Mexico

This iconic plant of the Southwest is a shrub. Small, green leaves appear off and on throughout the year along thorny, brown stems called canes, often responding to rain or increased humidity. Vibrant red/orange flowers appear in spring that attract hummingbirds. Ocotillo aren't fast-growers, and they need patience when growing in the landscape. They are available as bare root and nursery-grown. Bare root ocotillo doesn't always transplant successfully. If possible, choose an ocotillo grown in a nursery container with a much higher success rate. Plant ocotillo in full sun and remove any wires surrounding the canes. Water deeply once a month, spring through fall, to promote leafing out and blooms.

CACTI/SUCCULENT PROFILES

Hesperaloe parviflora

RED YUCCA

Bloom Period: Spring, Summer

Mature Height x Spread: 3 × 5 ft (91 cm × 1.5 m)

Exposure: Full sun, reflected heat, part shade

Cold Tolerance: -20°F (-29°C)

Origin: Texas, Mexico

This popular succulent has dark green leaves that resemble ornamental grass in shape. Coral-colored flowers appear in spring and throughout the warm season, attracting hummingbirds. Placing them in the landscape allows enough room to grow to their mature width. Red yucca looks best planted near large rocks scattered throughout the landscape. There are other varieties of red yucca, including one with light-yellow flowers and 'Brakelights', which has dark red flowers and is slightly smaller in overall size. Giant hesperaloe (*H. funifera*) is a related species with large, bright green leaves. Prune away flowering stalks to the base once they fade. Don't prune the top of the leaves, which leaves the plant ugly, and it takes a few years to recover.

Opuntia basilaris

BEAVERTAIL PRICKLY PEAR

Bloom Period: Spring

Mature Height x Spread: 2 × 3 ft (61 × 91 cm)

Exposure: Full sun

Cold Tolerance: 0°F (-18°C)

Origin: Southwest North America

The petite size of this prickly pear species makes it a great selection for smaller spaces. While many other prickly pear cacti can become unwieldy and large, beavertail stays neat and compact. Its name is because its pads resemble a beaver's tail. Glochids cover each pad, so take care to avoid touching them directly. Bright pink blooms cover the entire cactus in late spring, creating a stunning display. Plant in groups of three or more to maximize its impact or plant singly next to a large rock. Keep several feet away from areas with foot traffic.

CACTI/SUCCULENT PROFILES

Opuntia cacanapa 'Ellisiana'

HARDY SPINELESS PRICKLY PEAR

Bloom Period: Spring, Summer
Mature Height x Spread: 3–4 × 6 ft (91 cm–1.2 m × 1.8 m)
Exposure: Full sun
Cold Tolerance: 0°F (-18°C)
Origin: Texas, Mexico

The flat pads of this cold-tolerant prickly pear are brought into sharper focus by its relatively smooth surface. This cactus is quite versatile in the landscape as it lacks glochids and spines, increasing the areas where it can be used. Unlike the smooth Indian fig (*Opuntia ficus-indica*), which gets very large and may need frequent pruning, this species doesn't get as large. In late spring, pale yellow flowers appear at the tips of the pads, which are followed by edible, red fruit. Plant near agave or other succulents with a spreading growth habit to create a nice contrast of shapes and textures. Alternatively, pair with low-water flowering groundcovers.

Opuntia santa-rita

SANTA RITA PRICKLY PEAR, PURPLE PRICKLY PEAR

Bloom Period: Spring
Mature Height x Spread: 4 × 6 ft (1.2 × 1.8 m)
Exposure: Full sun
Cold Tolerance: 15°F (-9°C)
Origin: Southwest United States, Mexico

The attractive blue-gray color of this cactus stands apart from other species of prickly pear due to the shades of purple coloring that are more pronounced in cold weather. The stems are paddle-shaped and covered in tiny hairlike bristles called *glochids*, along with a few longer spines. In late spring, large yellow flowers appear, which contrast beautifully with the cool color of the pads. Edible fruits follow the flowers. Use Santa Rita prickly pear near yellow-flowering perennials and shrubs to emphasize its color. As with most species of prickly pear, keep away from high-traffic areas as its glochids and spines are extremely irritating to the skin. Cochineal scale is an insect that can produce cottony masses on the pads and is controlled by spraying it off with a hose.

CACTI/SUCCULENT PROFILES

Pachycereus marginatus
(syn. Marginatocereus marginatus)

MEXICAN FENCE POST

Bloom Period: Spring

Mature Height x Spread: 10 × 6 ft (3 × 1.8 m)

Exposure: Full sun

Cold Tolerance: 25°F (-4°C)

Origin: Mexico

The dark green, upright stems of this cacti are a favorite of landscape designers who use its relatively straight growth habit to great effect along walls. Delineated white margins run up and down the ribs of Mexican fence post cacti, creating a nice contrast in color. Short spines occur only on the white margins leaving the areas in-between smooth. This cactus has a relatively moderate growth rate, with new stems emerging from the base. Mexican fence post can be used in place of shrubs to create a tall, succulent border or is equally at home when planted as a single specimen. Its versatility extends to container gardening, and it is easy to start from cuttings. Provide supplemental irrigation spring through fall in low desert gardens.

Pachycereus schottii f. monstrosus
(syn. Lophocereus schottii)

TOTEM POLE

Bloom Period: None of note

Mature Height x Spread: 8–10 × 3–6 ft (2.4–3 m × 91 cm–1.8 m)

Exposure: Full sun

Cold Tolerance: 25°F (-4°C)

Origin: Mexico

Totem pole 'monstrosus' is a mutated form of an upright, thorny cactus (*Pachycereus schottii*). This mutant cactus has two characteristics that make it a big favorite in the arid landscape. First, its succulent, gray-green stems are lumpy, resembling a drippy candle. Another desirable feature is that it is thornless, making it an excellent choice for areas where people walk, such as a courtyard, patio, or sidewalk. It will branch from the bottom and occasionally higher up, producing additional "arms." *Pachycereus schottii f. mieckleyanus* is another form of this cactus that has thinner stems and fewer lumps. Whichever form you use, they both are an excellent option for narrow planting areas near pools or planting in containers. This unique cactus is easily propagated by cuttings. Plant in well-drained soil. Water monthly in the summer.

CACTI/SUCCULENT PROFILES

Portulacaria afra

ELEPHANT'S FOOD

Bloom Period: None of note

Mature Height x Spread: 2–3 × 2–4 ft (61–91 cm × 61 cm–1.2 m)

Exposure: Full sun, part shade, shade

Cold Tolerance: 25°F (-4°C)

Origin: South Africa

This trailing succulent has bright green, succulent leaves and maroon stems. Elephant's food is very easy to grow and adds a nice splash of green in areas with full sun and shade, making it highly versatile in the dry climate garden. It makes an excellent container plant and can be planted in a colorful pot to brighten up a shady or sunny spot. Its spreading growth habit makes it suitable as a low-water groundcover. Other cultivars are available with variegated foliage, such as 'Aurea' and 'Mediopicta'. 'Elephant Bush' is a form that grows much larger, and 'Prostrata' is a dwarf cultivar. Elephant's food does require supplemental water and should be put on a watering schedule for succulents. Little maintenance is needed. Prune in spring to remove frost-damaged growth and keep it at the desired size for its space.

Trichocereus hybrid

HYBRID TORCH CACTUS

Bloom Spring, early Summer

Mature Height x Spread: 1–2 × 3–4 ft (30–61 cm × 91 cm–1.2 m)

Exposure: Full sun

Cold Tolerance: 0°F (-18°C)

Origin: Argentina

In spring, large flowers cover the spiny stems of this South American native, creating a jaw-dropping display. Flower color depends on the hybrid and ranges from white, pale pink, magenta, red, and yellow. Popular hybrids include 'Blood' with red flowers, 'First Light' with pale-pink blooms, and 'Flying Saucer', with a blend of pink and orange. The size of the blooms range from 2 to 8 inches (5–20 cm). Many hybrids will bloom more than once. The cacti are unremarkable and fade into the background when not in flower. Fertilize with a slow-release fertilizer in spring. Provide afternoon shade or filtered sunlight in low-desert gardens. Protect with shade cloth during their first summer. They do best with bimonthly water when temperatures are above 80°F (27°C).

GRASS AND PALM PROFILES

Bouteloua gracilis

BLUE GRAMA

Bloom Period: Spring, Summer

Mature Height x Spread: 2 × 2 ft (61 × 61 cm)

Exposure: Full sun

Cold Tolerance: -30°F (-34°C)

Origin: North America

This is a lovely grass with medium-green leaves with an upright, arching shape. The foliage is very thin, creating an airy appearance within the garden. In summer, cream-colored flowers are produced that look like eyebrows, giving a whimsical look to this grass. The flowers last through the rest of the year. Plant blue grama wherever you want to introduce green softness to bare spots. It makes a natural pairing for large rocks but can also be used in straight rows along pathways or against structures. 'Blonde Ambition' is a popular variety. Prune once a year by shearing it back to the ground in spring once weather begins to warm.

Muhlenbergia capillaris

PINK MUHLY

Bloom Period: Fall

Mature Height x Spread: 3 × 3 ft (91 × 91 cm)

Exposure: Full sun

Cold Tolerance: 0°F (-18°C)

Origin: Southeastern United States

This is a finely textured ornamental grass with bright green foliage. It has a lovely arching growth habit. A cloud of tiny burgundy plumes is produced in the fall, adding cool-season interest to the garden. As the weather cools in winter, the plumes fade to an attractive wheat color. In spring, new green growth appears. Use pink muhly against a wall or as a border. Scatter several grasses, boulders, and succulents such as agave for great texture contrast in a more natural arrangement. Shear back to the ground in early spring.

Muhlenbergia rigens

DEER GRASS

Bloom Period: Fall

Mature Height x Spread: 4–5 × 3–5 ft (1.2–1.5 m × 91 cm–1.5 m)

Exposure: Full sun, reflected heat, part shade

Cold Tolerance: -10°F (-23°C)

Origin: Southwestern United States

The gray-green foliage of this ornamental grass adds a welcome lushness to the dry climate landscape. In fall, spikes of tan flowers create additional height and interest. Deer grass is easy to grow and makes an excellent background or filler plant throughout the landscape. It's also suitable for use in mass, such as underneath a tree that allows filtered sunlight through its branches. Prune back to the ground in spring once the threat of frost has passed. No other pruning is required.

GRASS AND PALM PROFILES

Brahea armata

MEXICAN BLUE FAN PALM

Bloom Period: Summer through Fall

Mature Height x Spread: 20–30 × 10 ft
(6.1–9.1 × 3 m)

Exposure: Full sun

Cold Tolerance: 15°F (-9°C)

Origin: Mexico

The prominent fan-shaped fronds of this Mexican native are prized for their blue-gray color. This large palm is slow-growing, so purchasing a larger specimen is worthwhile. Plant where it has room to grow to its mature size. Use it to visually anchor the corners of a property or where it is the vertical focal point of the landscape. The flowering plumes of Mexican blue fan palm are pretty showy, adding a cream-colored accent followed by black fruit. In late summer/early fall, prune away palm fronds that are significantly drooping. Fruiting plumes can be removed at any time. Take special care not to over-prune as it may not recover. Apply a palm fertilizer in early summer.

Chamaerops humilis

MEDITERRANEAN FAN PALM

Bloom Period: Spring

Mature Height x Spread: 15–20 × 15–20 ft
(4.6–6.1 × 4.6–6.1 m)

Exposure: Full sun, reflected heat, part shade

Cold Tolerance: 5°F (-15°C)

Origin: Mediterranean region

Fan-shaped fronds grace the multiple trunks of this medium-sized palm tree. The bright green color of the foliage and clumping growth habit make this palm a big favorite in both arid and colder, wetter climates. Mediterranean fan palm is an excellent substitute for larger palm trees as it doesn't overwhelm the landscape with its size. Plant near pools, a bare wall, or a garden bed where its pleasing shape will be appreciated. Small, pale yellow flowering stalks are produced followed by black fruit. Prune in late summer/early fall, removing palm fronds that droop and turn yellow. The fruiting branch can be removed at this time as well. Remove any new unwanted trunks as needed. Fertilize in early summer with a palm tree fertilizer.

Phoenix roebelenii

PYGMY DATE PALM

Bloom Period: Winter, Spring

Mature Height x Spread: 6–10 × 6–8 ft
(1.8–3 × 1.8–2.4 m)

Exposure: Full sun, part shade

Cold Tolerance: 25°F (-4°C)

Origin: South Asia

This small palm has dark green fronds that reach approximately 3–4 feet (91 cm–1.2 m) in length. The featherlike leaves provide a nice arching growth habit. Pygmy date palms are commonly grown as multi-trunk palms but can be trained into a single trunk. In spring, small, yellow flowers turn into clusters of black date fruit. Use where you want to add an oasis-like feel near a pool or patio. Avoid planting in areas with reflected heat, particularly in low-desert climates where afternoon shade is preferred. Limit trunks to three or five and prune away the rest at the base as they appear. Remove palm fronds as they turn yellow and brown in late summer or early fall. Fertilize in early summer with palm fertilizer.

White globe mallow shrubs (*Sphaeralcea ambigua*) thrive despite experiencing reflected heat from the wall behind them and the sidewalk in front.

WHAT IS REFLECTED HEAT, AND WHY IS IT A BIG DEAL?

We have talked about the stresses that living in a hot, dry climate puts on plants—from low humidity, the intensity of the sun, and the heat, which can be extreme. However, there are places within the landscape around your house where surrounding structures and hard surfaces amplify these challenges.

If you have ever seen the term *reflected heat*, it is pretty descriptive of this situation that plants need to deal with in specific locations. This term describes areas where walls, pavement, sidewalk, or a street absorbs the sun's rays and heats up. The heat is then radiated back out into the atmosphere and onto things nearby—namely, plants. So, imagine a plant that already deals with the stress of growing in full sun and amplifies that by having even more heat reflecting from the sidewalk or wall just a few feet away. Talk about a challenging place to grow!

Not surprisingly, many plants aren't equipped to handle this extra stress. Yes, they may be fine growing in full sun in the middle of the garden, but when put in an area with reflected heat, all bets are off. The good news is that many shrubs, groundcovers, vines, and succulents survive in these locations and thrive.

WHERE ARE REFLECTED HEAT LOCATIONS FOUND IN THE GARDEN?

To grow plants successfully in a hot, arid climate, we need to identify where reflected heat occurs around our house. We then use the right plant adapted to tolerate the challenges of intense heat and sun.

Areas with reflected heat are almost always associated with full sun. Hard surfaces such as walls, sidewalks, and streets absorb the sun's rays and re-radiate them back toward the plant. You will typically find spots with reflected heat on the corners of your driveway, areas 2 to 4 feet (61 cm–1.2 m) from the sidewalk or street, and south- or west-facing walls. When you think about locations in your landscape where plants typically struggle in the summer months, it is likely in one of these hot spots.

It's essential to include plants tolerant of stressful conditions like reflected heat to create a garden that is resilient to a changing climate with increasing temperatures and periods of drought. In the plant lists found earlier in this section, I have noted recommended shrubs, groundcovers, vines, and succulents suitable for reflected heat situations. However, I have listed more plants that will flourish and add beauty to the surrounding area on the following chart.

The groundcover damianita (*Chrysactinia mexicana*) adds beauty alongside a driveway where the effects of sun and heat are amplified.

10 MORE PLANTS FOR FULL SUN & REFLECTED HEAT

COMMON NAME	BOTANICAL NAME	MATURE SIZE HxW	COLD HARDINESS	NOTES
Bougainvillea	*Bougainvillea spp.*	5–10 × 6–12 ft (1.5–3 × 1.8–3.7 m)	25°F (-4°C)	Tropical shrub or upright, shrubby vine. Colorful brachts (usually dark pink) appear year-round in frost-free weather. Has lush green foliage and thorns.
Woolly Butterfly Bush	*Buddleja marrubifolia*	5 × 5 ft (1.5 × 1.5 m)	10°F (-12°C)	Soft gray foliage, orange flowers appear off and on throughout the year. Attracts butterflies.
Toothless Desert Spoon	*Dasylirion quadrangulatum*	6 × 6 ft (1.8 × 1.8 m)	10°F (-12°C)	Large succulent with narrow spiky, bright green leaves that fan outward. Leaves have serrated edges with tiny "teeth."
San Marcos Hibiscus	*Gossypium harknessii*	3 × 5 ft (91 cm × 1.5 m)	20°F (-7°C)	Large shrub with heart-shaped, medium green leaves. Yellow flowers with red centers appear off and on spring through fall.
Giant Hesperaloe	*Hesperaloe funifera*	6 × 6–8 ft (1.8 × 1.8–2.4 m)	-10°F (-23°C)	Spiky accent plant that resembles a large ornamental grass. A tall flowering stalk appears in summer on mature plants.
Violet Silverleaf	*Leucophyllum candidum*	3 × 3 ft (91 × 91 cm)	10°F (-12°C)	Species of Texas sage with small gray leaves. Deep purple blooms appear intermittently summer through fall in response to increased humidity.
Palo Blanco	*Mariosousa willardiana* syn. *Acacia willardiana*	20 × 15 ft (4.6 × 6.1 m)	20°F (-7°C)	Small tree with distinctive white bark and airy foliage with a drooping growth habit. Ideal for narrow spots where a tree is desired.
Bear Grass	*Nolina microcarpa*	3–5 × 4–7 ft (91 cm–1.5 m × 1.2.–2.1 m)	-10°F (-23°C)	Grass-like plant with medium green leaves that form light tan curlicues at the tips. May produce tall flowering stalks in summer.
Jojoba	*Simmondsia chinensis*	8 × 10 ft (2.4 × 3 m)	10°F (-12°C)	Oval-shaped, gray-green leaves. Cosmetics are made from the wax of the female plant's small brown fruit.
Arizona Rosewood	*Vauquelinia californica*	15 × 10 ft (4.6 × 3 m)	-10°F (-23°C)	Deep green foliage. Creamy white flowers appear in late spring/early summer. Good substitute for oleander.

FINDING PLANTS THAT LIKE BOTH HEAT AND SHADE

Within arid and semi-arid regions, most native plants favor sunny growing conditions. It's much easier to find shrubs, groundcovers, and succulents that prefer sun than those that like shade. Most plants that thrive in hot, dry climate regions do so in full sun. Therefore, when selecting plants for shade, we need to use plants that aren't only adapted to less light but can also tolerate our hot summer temperatures.

There is shade present in most gardens, and many plants don't do well in areas without much sun. As a result, plants can have a sprawling growth habit as their stems and branches attempt to grow toward the sun, which is rather unattractive. In addition, many plants won't bloom in low light conditions—this is especially true of flowering annuals. However, shade doesn't have to be a limiting element in your garden as there are attractive plant options.

THE DIFFERENT TYPES OF SHADE

There are two types of shade where it's relatively easy to grow plants—bright shade and filtered shade. Areas with bright shade have high light levels even though they may not receive any direct sun. These tend to be located around your home near the patio and front entry. Filtered shade (also referred to as part shade and filtered sunlight) are areas near trees whose branches filter out some of the sun but not all. You will find these spots near the outer branches of traditional shade trees or underneath the wispy foliage of acacia, mesquite, and palo verde tree species.

Deep shade is a problematic exposure for plants to grow in. If you have a deep entry leading to your front door, a north-facing patio, or an atrium, you have likely struggled to find plants to grow in these areas. I recommend adding colorful or decorative containers planted with shade-loving houseplants in locations like these, bringing them indoors during cold winters. The beauty of the container is important in these situations as they help add a decorative element when lack of plant choice and blooms are factors.

I've listed a number of plants that do well in light or part shade conditions within the plant profiles. However, there are more that you may want to consider adding that I've listed in the chart on page 185. The good news is that shade doesn't have to be a limiting element in your garden with these attractive plant options.

The orange flowers of coral aloe (*Aloe striata*) add cool-season interest and attract hummingbirds.

Red flowers add vibrant color to the lush green leaves of this crown-of-thorns (*Euphorbia milii*) succulent.

10 MORE PLANTS FOR SHADE

COMMON NAME	BOTANICAL NAME	MATURE SIZE HxW	COLD HARDINESS	NOTES
Coral Aloe	*Aloe striata*	2 × 2 ft (61 × 61 cm)	25°F (-4°C)	Light green succulent with clumping leaves. Vibrant orange flowers appear on 2-foot-tall stems late winter into spring.
Foxtail Asparagus Fern	*Asparagus densiflorus* 'Myers'	2 × 2 ft (61 × 61 cm)	25°F (-4°C)	Attractive perennial with bright green plumes and small narrow leaves with a fernlike appearance.
Bulbine	*Bulbine frutescens*	1 × 2 ft (30 × 61 cm)	10°F (-12°C)	Great in containers or in the ground. Groundcover with light green, grass-like succulent leaves. Yellow or orange flowers appear fall to spring.
Sago Palm	*Cycas revoluta*	6 × 4–5 ft (1.8 × 1.2–1.5 m)	15°F (-9°C)	Not a true palm but a cycad. The deep green foliage resembles palm fronds. Slow growing and suitable for large pots or in ground. Toxic if ingested.
Crown-of-Thorns	*Euphorbia milii*	2–3 × 3–4 ft (61–91 cm × 91 cm–1.2 m)	30°F (-1°C)	Bright green leaves cover thorny stems. Red (or pink) flowers appear sporadically throughout the year. Milky sap is irritating to skin and eyes.
Creeping Fig	*Ficus pumila*	20 × 20 ft (6.1 × 6.1 m)	10°F (-12°C)	Slow-growing, lush green vine that clings to surfaces. Leaves on mature vines are three times larger than juvenile ones.
Heavenly Bamboo	*Nandina domestica*	2–6 × 5 ft (61 cm–1.8 m × 1.5 m)	10°F (-12°C)	Bright green foliage that resembles bamboo is tinged red in winter. Flowers in spring, and red berries appear in fall/winter.
Arabian Jasmine	*Sambac jasminoides*	3–5 × 4–5 ft (91 cm–1.5 m × 1.2–1.5 m)	25°F (-4°C)	Lush green shrub with tropical look. Fragrant, white flowers appear mid-spring into summer.
Yellow Dot	*Sphagneticola trilobata* (*syn. Wedelia trilobata*)	2 × 6 ft (61 cm × 1.8 m)	30°F (-1°C)	Fast-growing groundcover with dark green foliage. Yellow flowers appear in spring and summer.
Purple Heart	*Tradescantia pallida*	15 × 10 ft (4.6 × 6.1 m)	10°F (-12°C)	Succulent groundcover with deep purple foliage. Flowers appear in summer.

Little leaf cordia (*Cordia parvifolia*) produces papery-white blooms due to increased humidity and rain.

Pink fairy duster (*Calliandra eriophylla*) adds beauty to the desert garden while attracting hummingbirds in spring.

THE QUEST FOR LOWER WATER USE PLANTS

The water supply in many dry climate regions is shrinking. According to experts and our observations, we are in a mega-drought with no end in sight. In many communities, water restrictions are already in place that limit what days and times you can water your plants and are likely to spread to other municipalities. Consequently, the interest in creating a landscape filled with water-thrifty plants increases.

Considering that an average of 70 percent of household water use goes to outdoor use, it makes sense to look at your plants' water needs. First, see how often you water your plants and whether you are using the most efficient irrigation method. The next step is to determine if you have plants that need a lot of water—especially those that are native to regions that have higher levels of humidity and receive more rainfall. Finally, consider switching out these high-water-use plants for those that need less.

LOW WATER USE SHRUBS

The first type of plants to look at adding are those native to your region. Once established, they will need the least amount of maintenance and will grow with minimal supplemental water. If you want to expand your plant choices, check out plants from dry, arid regions similar to yours.

Note that the term "drought-tolerant" can be misleading. A plant considered drought-tolerant from another type of climate may need regular water in an arid region. Additionally, a plant that is called drought-tolerant doesn't mean that it needs no supplemental water. Overall, a drought-tolerant plant doesn't use excessive amounts of supplemental water when grown under similar conditions in its native region.

It's important to note that even highly drought-tolerant plants will need to be watered at regular intervals to become established through the first year after planting to grow a healthy root system. With increasing temperatures and lower rainfall, even native plants can struggle without supplemental water, so it is essential to watch your plants for signs of underwatering, such as smaller leaves, lack of blooms, and leaves dropping to the ground. With cacti and other succulents, look for wrinkled leaves or deeper folds with columnar cacti indicating they need water.

In the chart on page 188, I've listed water-wise shrubs that can be used with large and smaller cacti and other succulents to create an attractive landscape that doesn't require much supplemental water.

10 MORE HIGHLY DROUGHT-TOLERANT PLANTS

COMMON NAME	BOTANICAL NAME	MATURE SIZE HxW	COLD HARDINESS	NOTES
Goldeneye	*Bahiopsis parishii* (syn. *Viguieria deltoidei*)	3 × 3 ft (91 × 91 cm)	15°F (-9°C)	Naturally rounded shrub with medium-green foliage. Yellow daisy-like flowers appear in spring and sometimes in fall. Use in full sun.
Pink Fairy Duster	*Calliandra eriophylla*	3 × 3 ft (91 × 91 cm)	10°F (-12°C)	Naturally rounded shrub with small green leaves. Produces pink puffball blooms in spring. Attracts hummingbirds. Needs full sun.
Little Leaf Cordia	*Cordia parvifolia*	5–6 × 8 ft (1.5–1.8 × 2.4 m)	10°F (-12°C)	Semi-evergreen shrub with gray-green foliage. White flowers appear in summer, often in response to humidity. Plant in full sun.
Brittlebush	*Encelia farinosa*	3 × 4 ft (91 cm × 1.2 m)	10°F (-12°C)	Naturally rounded shrub with gray-green foliage. Yellow, daisy-like flowers appear in spring. Self-seeds. Requires full sun.
Turpentine Bush	*Ericameria laricifolia*	3 × 3 ft (91 × 91 cm)	-10°F (-23°C)	Lush green, aromatic foliage with narrow, needle-shaped leaves. Golden yellow flowers appear in fall. Needs full sun.
Flattop Buckwheat	*Eriogonum fasciculatum v. polifolium*	1 × 2 ft (30 × 61 cm)	10°F (-12°C)	Dusty-green, narrow leaves on low-growing mounded shrub. Clusters of tiny pale pink/white flowers appear in spring and summer. Plant in full sun.
Desert Lavender	*Hyptis emoryi*	6–10 × 6–12 ft (1.8–3 × 1.8–3.7 m)	15°F (-9°C)	Large shrub with aromatic gray foliage with small lavender flowers that appear spring through fall. Requires full sun.
Creosote	*Larrea tridentata*	6 × 8 ft (1.8 × 2.4 m)	0°F (-18°C)	Slow-growing shrub with aromatic foliage that smells like rain. Tiny yellow flowers appear in spring. Needs full sun.
Jojoba	*Simmondsia chinensis*	6–8 × 8–10 ft (1.8–2.4 × 2.4–3 m)	15°F (-9°C)	Attractive gray-green oval leaves with a leathery texture. Grows slowly. Female shrubs produce hard, brown fruit. Best in full sun.
Arizona Rosewood	*Vauquelinia californica*	15 × 10 ft (4.6 × 3 m)	-10°F (-23°C)	Deep green foliage. Creamy white flowers appear in late spring/early summer. Good substitute for oleander. Plant in full sun.

Vibrant blooms cover the deep green foliage of sandpaper verbena (*Glandularia rigida*).

Blanket flower (*Gaillardia x grandiflora*) adds sunny colors to spring and summer gardens.

USING PLANTS TO DECORATE YOUR OUTDOOR SPACE

The easiest way to add interest to your landscape is by adding color. We do the same thing inside our homes by adding a vibrant color as an accent to attract attention. For outdoor spaces, most homeowners say they want more color. People are naturally attracted to color, which is an excellent way to up your home's curb appeal without spending much money.

Within arid and semi-arid regions, you'll find plants with varying shades of green, from lush, saturated greens to softer tones with a gray or bluish tint. Combining plants with different foliage colors creates a nice contrast and adds a lovely backdrop for colorful blooms. So whether your preference is for shades of red, orange, and yellow or cooler tones like purples and pinks, there are plenty of arid-adapted plants to choose.

When selecting plants for color, remember it's important to use color strategically, concentrating it in high-profile areas where it's likely to be viewed. These areas include the corners of your landscape, alongside your driveway, entry path, against walls, and near windows. A single groundcover or small shrub isn't likely to make a color impact, so mass smaller plants in amounts of three or five.

PLANTS FOR COLOR IMPACT

My natural garden style is one with many colorful flowering plants. However, I know how easy it is to go overboard with color leading to a busy or haphazard look. I recommend focusing on three primary colors for your plants. As you go through the plant profiles and this chart, look for plants in those colors. It's okay to add small pockets of additional color throughout the landscape or in one spot. To bring another color element to your garden, you can have three colors represented in your winter/spring garden and a different set of colors for your summer garden.

To maximize the length of time you have plants flowering throughout the year, select plants that bloom at different times. In addition, if you are pairing plants together to enjoy their blooms together, make sure that they bloom at the same time of year.

Shrubs and perennials are where we usually look to add color to the garden. I've listed many colorful options within the main plant profiles in this section, but here are more for you to consider.

10 MORE PLANTS FOR COLOR GARDENS

COMMON NAME	BOTANICAL NAME	MATURE SIZE HxW	COLD HARDINESS	NOTES
Desert Marigold	*Baileya multiradiata*	1 × 1 ft (30 × 30 cm)	-10°F (-23°C)	Short-lived perennial that self seeds. Yellow daisy flowers appear off and on throughout the year over gray foliage. Extremely drought tolerant.
Mexican Bird-of-Paradise	*Caesalpinia mexicana*	12–15 × 12 ft (3.7–4.6 × 3.7 m)	10°F (-12°C)	Shrub that is often trained into a small tree. Yellow flowers appear over lacy green foliage spring through summer.
Little John Bottlebrush	*Callistemon* 'Little John'	3 × 5 ft (91 cm × 1.5 m)	20°F (-7°C)	Medium shrub with evergreen foliage. Red, brush-like flowers appear in spring over narrow, gray-green leaves.
Skyflower	*Duranta erecta*	5–10 × 5–8 ft (1.5–3 × 15.–2.4 m)	28°F (-2°C)	Upright shrub with lush green, arching branches. Sprays of deep purple flowers appear in spring and summer.
Blanketflower	*Gaillardia x grandiflora*	1–2 × 1–2 ft (30–60 × 30–60 cm)	-40°F (-40°C)	Small perennial with green leaves and produces blooms in shades of orange, red, and yellow in spring and fall.
Sandpaper Verbena	*Glandularia rigida*	1 × 2 ft (30–60 cm)	10°F (-12°C)	Dusty-green, narrow leaves on low-growing mounded shrub. Clusters of tiny pale pink/white flowers appear in spring and summer.
Texas Ranger 'Lynn's Legacy'	*Leucophyllum langmaniae*	5 × 5 ft (1.5 × 1.5 m)	10°F (-12°C)	Large shrub with medium-green foliage. Produces purple blooms with increased humidity in warm season.
Mt. Lemmon Marigold	*Tagetes lemmonii*	3 × 3 ft (91 × 91 cm)	10°F (-12°C)	Aromatic, finely textured foliage that is medium green. Yellow, daisy-like flowers appear in fall.
Cape Honeysuckle	*Tecoma capensis*	15 × 10 ft (4.6 × 3 m)	-10°F (-12°C)	Shrub with attractive, deep green foliage. Can be trained upright to cover walls. Deep orange blooms appear spring through fall.
Bells of Fire™ Tecoma	*Tecoma x* 'Bells of Fire'	3–5 × 3–5 ft (91 cm–1.5 m × 91 cm–1.5 m)	20°F (-7°C)	Popular Tecoma hybrid with lush green foliage. Trumpet-shaped red flowers bloom spring through early fall.

A flagstone pathway winds through a garden with dry-climate plants that attract pollinators.

10 MORE PLANTS FOR POLLINATOR GARDENS

COMMON NAME	BOTANICAL NAME	MATURE SIZE HxW	COLD HARDINESS	NOTES
Flame Acanthus	*Anisacanthus quadrifidus v. wrightii*	3–4 × 3–4 ft (91 cm–1.2 m × 91 cm–1.2 m)	0°F (-18°C)	Small shrub with orange flowers in summer that attract butterflies and hummingbirds. Plant in sun, filtered shade.
Pine-Needle Milkweed	*Asclepias linearis*	2–3 × 2–3 ft (61–91 × 61–91 cm)	15°F (-9°C)	Perennial with narrow leaves. Creamy-white blooms appear spring to fall. Attracts butterflies. Plant in sun.
Desert Milkweed	*Asclepias subulata*	4 × 4 ft (1.2 × 1.2 m)	25°F (-4°C)	Succulent with upright stems. Pale yellow flowers attract bees and butterflies. Plant in sun.
Woolly Butterfly Bush	*Buddleja marrubifolia*	4–5 × 5 ft (1.2–1.5 × 1.5 m)	10°F (-12°C)	Shrub with velvety gray leaves. Round orange flowers attract bees and butterflies. Plant in sun.
Rock Verbena	*Glandularia pulchella*	1 × 3 ft (30 × 91 cm)	10°F (-12°C)	Bright green groundcover with purple flowers that attracts bees, butterflies, and hummingbirds. Plant in sun.
Passion Vine	*Passiflora foetida*	10 × 10 ft (3 × 3 m)	25°F (-4°C)	Lush green vine with purple-white flowers that attracts bees and butterflies. Plant in sun or shade.
Rock Penstemon	*Penstemon baccharifolius*	1 × 3 ft (30 × 91 cm)	0°F (-18°C)	Clumping perennial with orange-red flowers that attract bees and hummingbirds. Plant in sun.
Chaparral Sage	*Salvia clevelandii*	5 × 5 ft (1.5 × 1.5 m)	10°F (-12°C)	Shrub with aromatic leaves and purple blooms that attracts pollinators. Plant in full sun (afternoon shade in desert).
Mealy Cup Sage	*Salvia farinacea*	2 × 2 ft (61 × 61 cm)	10°F (-12°C)	Lush green shrub and violet flowers attracts butterflies and hummingbirds. Plant in sun (filtered shade in low desert).
Mexican Bush Sage	*Salvia leucantha*	3–4 × 3 ft (91 cm–1.2 m × 91 cm)	18°F (-8°C)	Shrubby perennial with purple/white flowers that attract bees, butterflies, and hummingbirds. Plant in sun (afternoon shade in low desert).

WHY YOU SHOULD ADD POLLINATOR PLANTS TO YOUR GARDEN

As urban development continues to encroach on the surrounding natural areas, plants that provide food for pollinators, such as bees, birds, and butterflies, are destroyed. As a result, pollinators have decreased due to habitat loss and a lack of plants necessary for survival. Yet, pollinators play an essential part in the food chain and plant survival as they help plants form seeds to grow a new generation of plants.

The good news is that we can play an essential part in helping pollinators by what we plant in our gardens! We select plants that will attract bees, birds, and butterflies to do this. Pollinator plants are those that provide a food source for these essential creatures. In the case of butterflies, we have two different types of plants needed to attract them—host (larval) plants where butterflies lay their eggs and caterpillars eat and nectar plants that attract adult butterflies. It's okay to let caterpillars eat the host plants. That is what is supposed to happen, and the host plant will grow back new leaves.

POLLINATOR PLANTS FOR ATTRACTING BEES, BUTTERFLIES, AND HUMMINGBIRDS
Hummingbirds are favorite visitors to gardens. While you can put out hummingbird feeders with a sugar-water mixture, better yet, plant a variety of plants with nectar-rich flowers. They usually favor colorful flowers with a trumpet shape perfect for their long beaks. In addition to nectar, hummingbirds eat many tiny insects, including mosquitos. Other bird species also help pollinate plants by spreading pollen between the flowers they visit.

Bees are dependent on the pollen and nectar they collect from flowers. While we think of bees, like European honey bees, as a collective living in hives, other types of bees are solitary and often live on their own, rarely sting, and play a massive part in pollinating native plants. Solitary bees are not aggressive and don't produce honey. Most don't have stingers, and those that do rarely sting. These solitary bees play a huge role in pollination.

POLLINATOR PLANTS FOR ATTRACTING BEES, BUTTERFLIES, AND HUMMINGBIRDS
Many plants within the plant profiles attract pollinators, which I've indicated. However, there are even more for you to consider adding to your outdoor space. A garden filled with pollinator plants is beautiful, allows you to view these crucial animals up close, and helps you know that you are doing your part to support their vital work.

ACKNOWLEDGMENTS

The book that you hold in your hands would not have been possible without the generosity of many people.

First, a big thank you goes out to my dear friends and fellow gardeners who understand the beauty that is possible to achieve in arid and semi-arid regions, Shawna Coronado and Andrea Whitely, who encouraged me while writing this book. To my fellow dry climate authors and educators, I am blessed to have many of you as friends and am honored to belong to this "tribe" of dry garden experts who understand my passion for helping people learn how to garden in hot and dry conditions. You all serve as constant inspiration to me.

I am truly grateful for those who have kindly allowed me to share photos of their gardens—especially Shawna Coronado, Cliff Douglas, and Janet Lennox Moyer. Thank you to my many clients, friends, and students who let me use pictures from their gardens! A special thank you goes out to the Desert Botanical Garden in Phoenix and Tucson Botanical Gardens. They graciously allowed me to use photos from their gardens that showcase the true beauty of an arid garden.

To my current and past clients, students, and followers: Your desire and questions on how to create your own garden that thrives in heat and low humidity were invaluable in helping me decide what I needed to include in this book—thank you.

My garden journey would never have started without my dad. When I was eleven years old, he gave me a small plot of land in our California backyard and let me plant whatever I wanted. It was because of him that my love for gardening was kindled. Oh, how I wish he were here to see this book! Thank you to my mother, who is and continues to be my biggest cheerleader throughout my garden career. To my sister (and assistant), Jennifer, 'AZ Plant Lady' wouldn't be anything without your help and encouragement. Thank you to my youngest sister, Grace, who helped me with the photos in the book and took pictures of my garden. Her professional photographer's eye truly captures the uniqueness of the dry climate garden. Thank you for the frequent gardening questions from my brother, Scott, which remind me how much people need help in the garden. To my five children, thank you for being my "assistants" in the garden while you were growing up. I love that two of you have gardens of your own now, and I hope the rest of you will enjoy gardens of your own someday.

Finally, to my wonderful husband, who never complains when I ask him to come with me to buy plants I don't need, to dig countless holes, and who has held my hand and supported me throughout my entire horticulture career—this book wouldn't exist without you.

PHOTOGRAPHY CREDITS

All photos by the author, except for the following:

Kelly Bergin: page 45

Tammy Bettinger: page 171 (right)

Carianne Campbell: pages 97, 192

Robert Colter: page 12 (right)

Linda Cornelison: page 110

Shawna Coronado: pages 107, 113, 118 (top)

Jennifer Humphrey: page 170 (right)

Rachele Johnson: page 134

Todd Johnson: pages 11, 63, 64, 160 (right), 172 (left)

Esther Smith: page 163 (right)

Shutterstock: page 115

Grace Stufkosky: pages 98, 100

Andrea Whitely: page 23

RESOURCES

ADDITIONAL READING

Arizona, Nevada & New Mexico Month-by-Month Gardening: What to Do Each Month to Have a Beautiful Garden All Year, Jacqueline Soule. Beverly, MA: Cool Springs Press, 2016.

Desert Landscape School: A Guide to Desert Landscaping and Maintenance, Edited by Luanna Vargas, Desert Botanical Garden Desert Landscape School. Phoenix, AZ: Desert Botanical Garden, 2017.

Gardening in the Desert: A Guide to Plant Selection & Care, Mary Irish. Tucson: University of Arizona Press, 2000.

Growing the Southwest Garden, Judith Phillips. Portland, OR: Timber Press, 2015.

Hot Color, Dry Garden: Inspiring Designs and Vibrant Plants for the Waterwise Gardener, Nan Sterman. Portland, OR: Timber Press, 2018.

Landscape Plants for Dry Regions, Warren Jones and Charles Sacamano. Tucson, AZ: Fisher Books, 2000.

Lawn Gone!: Low-Maintenance, Sustainable, Attractive Alternatives for Your Yard, Pam Penick. Berkeley, CA: 10 Speed Press, 2013.

Native Plants for Southwestern Landscapes, Judith Mielke. Austin: University of Texas Press, 1993.

Plants for Dry Climates: How to Select, Grow, and Enjoy, Mary Rose Duffield and Warren Jones. Cambridge, MA: Perseus Publishing, 2001.

The Water-Saving Garden: How to Grow a Gorgeous Garden With a Lot Less Water, Pam Penick. Berkeley, CA: Ten Speed Press, 2016.

Water-Wise Plants for the Southwest, Mary Irish, Joe Lamp'l, Judith Phillips, and Nan Sterman. Beverly, MA: Cool Springs Press, 2007.

BOTANIC GARDENS

ABQ BIOPARK
Albuquerque, NM, cabq.gov/artsculture/biopark

ARIZONA-SONORA DESERT MUSEUM
Tucson, AZ, desertmuseum.org

BOYCE THOMPSON ARBORETUM
Superior, AZ, btarboretum.org

DESERT BOTANICAL GARDEN
Phoenix, AZ, dbg.org

HUNTINGTON LIBRARY, ART MUSEUM, AND BOTANICAL GARDENS
Pasadena, CA, huntington.org

KEYSTONE HERITAGE PARK
El Paso, TX, keystoneheritagepark.com

KINGS PARK & BOTANIC GARDEN
Perth, Western Australia, bgpa.wa.gov.au/kings-park

THE LIVING DESERT ZOO / GARDENS
Palm Desert, CA, livingdesert.org

LOS ANGELES COUNTY ARBORETUM
Arcadia, CA, arboretum.org

RED BUTTE GARDEN
Salt Lake City, UT, redbuttegarden.org

RUTH BANCROFT GARDEN
Walnut Creek, CA, ruthbancroftgarden.org

SAN DIEGO BOTANIC GARDEN
Encinitas, CA, sdbgarden.org

SANTA BARBARA BOTANICAL GARDEN
Santa Barbara, CA, sbbg.org

SANTA FE BOTANICAL GARDEN
Santa Fe, NM, santafebotanicalgarden.org

SPRINGS PRESERVE
Las Vegas, NV, springspreserve.org

TOHONO CHUL GARDENS
Tucson, AZ, tohonochul.org

TUCSON BOTANICAL GARDENS
Tucson, AZ, tucsonbotanical.org

UNIVERSITY OF CALIFORNIA BOTANICAL GARDEN
Berkeley, CA, botanicalgarden.berkeley.edu

WATER CONSERVATION GARDEN PARK
West Jordan, UT, conservationgardenpark.org

ONLINE RESOURCES

ALBUQUERQUE, NM WATERING RECOMMENDATIONS

505outside.com/watering-recommendations

ARIZONA MUNICIPAL WATER USERS ASSOCIATION: PLANTS USE IT WISELY

amwua.org/plants

CALIFORNIA NATIVE PLANT SOCIETY

cnps.org

CHIHUAHUAN DESERT EDUCATION COALITION

chihuahuandesert.org

COACHELLA VALLEY WATER DISTRICT

cvwd.org

DEBRA LEE BALDWIN (SUCCULENTS)

debraleebaldwin.com

DESERT GARDENING 101 (AUTHOR'S WEBSITE)

azplantlady.com

EL PASO, TX WATER

epwater.org

HIGH COUNTRY GARDENS

highcountrygardens.com

INTERNATIONAL SOCIETY OF ARBORISTS

isa-arbor.com

LADYBIRD JOHNSON WILDFLOWER CENTER

wildflower.org/plants

METROPOLITAN WATER DISTRICT OF SOUTHERN CALIFORNIA

bewaterwise.com

MOUNTAIN STATES WHOLESALE NURSERY

mswn.com

SOUTHERN NEVADA WATER AUTHORITY

snwa.com

TREES ARE GOOD

treesaregood.org

TRUCKEE MEADOWS WATER AUTHORITY

tmwa.com

UTAH DIVISION OF WATER RESOURCES

conservewater.utah.gov

WATER USE IT WISELY - ARIZONA

wateruseitwisely.com/saving-water-outdoors

ABOUT THE AUTHOR

NOELLE JOHNSON is a horticulturist, landscape consultant, and instructor. Originally from Southern California, Noelle has made the desert of central Arizona her home for over thirty-five years. She has a bachelor's of science in plant biology with a concentration in urban horticulture from Arizona State University. Popularly known as "AZ Plant Lady," Noelle has been teaching and inspiring countless desert-dwellers to create, grow, and maintain beautiful gardens that thrive in a hot, dry climate for over twenty-five years with one-on-one clients, writing for local and national garden publications, and writing for her website, www.azplantlady.com. A retired certified arborist, Noelle is also a popular speaker, and her online classes have allowed her to spread her expertise and knowledge beyond Arizona to other dry climate regions. Noelle's goal is to dispel the myths of dry climate gardening and provide guidance in simple steps to provide people the knowledge to create, grow, and maintain a beautiful garden that thrives in a hot, dry climate.

INDEX